# MODERN CHROMATOGRAPHIC TECHNIQUES

By:

| **Dr. Bhavin B. Dhaduk** | **Dr. Khushal M. Kapadiya** |
|---|---|
| Assistant Professor, School of Science, R. K. University, Rajkot- 360020 Gujarat (India).  | Assistant Professor, School of Science, R. K. University, Rajkot- 360020 Gujarat (India).  |

# ABOUT AUTHORS

## Dr. Bhavin Dhaduk

**Dr. Bhavin Dhaduk** received his Master Degree (2011) and Ph.D. Degree (2016) in Analytical Pharmaceutical Chemistry from Saurashtra University, Rajkot (Gujarat- India). After masters he worked in INTAS pharmaceutical LTD., Ahmedabad as a Research Associate and after that he joined his Ph.D. His research work mainly focused on synthesis, method development and validation, ultrasonic, thermal and XRD studies of derivatives of bisphenol compounds. He is the author of more than 12 national and international journal articles. He worked as Assistant Professor & Head in Shree H. N. Shukla Science college in 2015. Now he is working as an Assistant Professor in School of Science, R.K. University and Rajkot.

# Dr. Khushal Kapadiya

**Dr. Khushal Kapadiya** received his Master Degree (2011) and Ph.D. Degree (2016) in Organo-Pharmaceutical Chemistry from Saurashtra University, Rajkot (Gujarat- India). After masters he worked in INTAS pharmaceutical LTD., Ahmedabad as a Research Associate and after that he joined his Ph.D. His research work mainly focused on modeling and design of diverse scaffolds particularly MCRs and coupling reaction based chemistry. He is the author of more than 15 national and international journal articles. He joined School of Science, R.K. University, Rajkot, as an Assistant Professor in 2016.

# PREFACE

Chromatography is a powerful separation technique that is used in all branches of science and is often the only means of separating components from complex mixtures. The Russian Botanist M. Tswett coined the term chromatography in 1906. The first analytical use of chromatography was described by James and Martin in 1952, for the use of gas chromatography for the analysis of fatty acids mixtures. Traditional chromatographic techniques must now be rained because it does not satisfied all expectations and standards. Due to globalization, arrival of new techniques, recent information is now required. As a consequence, the versatilities of some processes associates with the ethical concerns. Therefore, in this book we described recent chromatographic techniques and its applications in different industries. There is a target to summarize basic knowledge of chromatographic techniques which is mostly applicable in current trends, so we mainly focused on some advanced topics. This book is mainly used for master as well as bachelor students which can summarize applications and uses of various types of unit processes and can learn proper theory.

We hope this book will be useful as a reference and will be a meaningful addition for advanced undergraduate and graduate students in Analytical Chemistry.

**Dr. Bhavin Dhaduk**
&
**Dr. Khushal Kapadiya**

## :::::::::::::: TABLE OF CONTENT::::::::::::::

| Sr. No. | Title of Content | Page No. |
|---|---|---|
| Chapter 1 : Chromatography : Basic concept and its application ||| 
| 1.0 | General Introduction | Page No. 1 |
| 1.1 | Principle of Chromatography | Page No. 1 |
| 1.2 | Classification of Chromatography | Page No. 2 |
| 1.3 | Distribution Coefficient (K) | Page No. 3 |
| 1.4 | Chromatographic Terms | Page No. 4 |
| 1.4.1 | Chromatogram | Page No. 4 |
| 1.4.2 | Retention Time ($t_R$) | Page No. 5 |
| 1.4.3 | Retention Volume ($V_R$) | Page No. 5 |
| 1.4.4 | Baseline Width (w) | Page No. 5 |
| 1.4.5 | Void time or void volume ($t_M$) | Page No. 5 |
| 1.4.6 | Chromatographic Resolution (Rs) | Page No. 6 |
| 1.4.7 | Retention Factor (k) | Page No. 7 |
| 1.4.8 | Column Selectivity (α) | Page No. 9 |
| 1.5 | Plate Theory (Efficiency of column) | Page No. 9 |
| 1.5.1 | Calculation of Chromatographic peak width | Page No. 11 |
| 1.6 | Rate theory (Band Broadening) | Page No. 12 |
| 1.6.1 | Eddy Diffusion(Multipath Diffusion) | Page No. 13 |
| 1.6.2 | Longitudinal diffusion | Page No. 13 |
| 1.6.3 | Mass Transfer | Page No. 14 |
| Chapter – 2 High Performance Liquid Chromatography |||
| 2.1 | Introduction | Page No. 15 |
| 2.2 | Why it is superior to traditional chromatography? | Page No. 15 |
| 2.3 | Principle | Page No. 16 |
| 2.4 | Column packing (stationary Phase) | Page No. 16 |
| 2.5 | Bonded phase columwn (Silanization) | Page No. 17 |
| 2.6 | Types of HPLC or Separation modes | Page No. 18 |

| 2.6.1 | Normal phase chromatography | Page No. 19 |
|---|---|---|
| 2.6.2 | Reverse phase chromatography | Page No. 19 |
| 2.7 | Elution modes | Page No. 20 |
| 2.7.1 | Isocratic elution | Page No. 20 |
| 2.7.2 | Gradient elution | Page No. 21 |
| 2.8 | HPLC Instrumentation | Page No. 21 |
| 2.8.1 | Working Function | Page No. 22 |
| 2.9 | Choice of solvent or mobile phase preparation | Page No. 23 |
| 2.9.1 | Solvent miscibility | Page No. 23 |
| 2.9.2 | Viscosity | Page No. 23 |
| 2.9.3 | UV cut-off | Page No. 24 |
| 2.9.4 | Polarity | Page No. 24 |
| 2.9.5 | Selectivity factor | Page No. 25 |
| 2.10 | Importance of mobile phase degassing | Page No. 28 |
| 2.11 | HPLC Pumping | Page No. 29 |
| 2.12 | Sample injection technique | Page No. 30 |
| 2.13 | HPLC column | Page No. 31 |
| 2.14 | HPLC Detectors | Page No. 32 |
| 2.14.1 | UV-visible and florescence detector | Page No. 33 |
| 2.14.2 | Photo diode array detector | Page No. 34 |
| 2.14.3 | Refractive index detector | Page No. 34 |
| 2.14.4 | Fluorescence Detector | Page No. 35 |
| Chapter – 3 Liquid chromatography-mass spectrometry (LC-MS) | | |
| 3.1 | Introduction | Page No. 37 |
| 3.2 | Importance of MS in LC or Sensitivy and Selectivity | Page No. 37 |
| 3.3 | Instrumentation | Page No. 38 |
| 3.4 | Interfacing LC and MS | Page No. 39 |
| 3.5 | Ion sources | Page No. 39 |
| 3.5.1 | API-ES (Electrospray ionization) | Page No. 40 |

| 3.5.2 | APCI (Atmospheric Pressure Chemical ionization) | Page No. 41 |
|---|---|---|
| 3.5.3 | APPI (Atmospheric Pressure photo ionization) | Page No. 42 |
| 3.6 | Mass Analysers | Page No. 43 |
| 3.6.1 | Quadrupole Mass analyser | Page No. 43 |
| 3.6.2 | Time of flight (TOF) Mass analyser | Page No. 45 |
| 3.6.3 | Ion Trap Mass analyser | Page No. 47 |
| 3.7 | Application | Page No. 47 |
| 3.7.1 | Molecular weight determination | Page No. 47 |
| 3.7.2 | Pharmaceutical Applications | Page No. 48 |
| 3.7.3 | Food applications | Page No. 49 |
| **Chapter – 4 Gas Chromatography (GC)** | | |
| 4.1 | Introduction | Page No. 51 |
| 4.2 | Difference of GSC and GLC | Page No. 51 |
| 4.3 | Instrumentation | Page No. 52 |
| 4.3.1 | Carrier gas system | Page No. 52 |
| 4.3.2 | Flow regulators and flow meter | Page No. 52 |
| 4.3.3 | Sample injection system | Page No. 53 |
| 4.3.4 | Column Configurations | Page No. 55 |
| 4.3.5 | Column temperature (temperature programming) | Page No. 57 |
| 4.4 | Detection system | Page No. 58 |
| 4.4.1 | Flame Ionization Detector (FID) | Page No. 58 |
| 4.4.2 | Thermal Conductivity Detector (TCD) | Page No. 59 |
| 4.4.3 | Electron Capture Detector (ECD) | Page No. 60 |
| **Chapter – 5 Gas chromatography-mass spectrometry (GC-MS)** | | |
| 5.1 | Introduction | Page No. 61 |
| 5.2 | Instrument | Page No. 61 |
| 5.3 | Sample used | Page No. 62 |
| 5.3.1 | Sampling techniques | Page No. 62 |
| 5.4 | Interface or coupling of GC-MS | Page No. 63 |
| 5.4.1 | Jet separator | Page No. 63 |

| | | |
|---|---|---|
| 5.4.2 | The Bieman concentrator | Page No. 65 |
| 5.4.3 | Direct introduction | Page No. 65 |
| 5.5 | Ion sources | Page No. 66 |
| 5.5.1 | Electron impact ionization (EI) | Page No. 66 |
| 5.5.2 | Chemical ionization (CI) | Page No. 67 |
| 5.6 | Analysers | Page No. 69 |
| 5.7 | Difference between LC-MS and GC-MS | Page No. 69 |
| 5.8 | Strength and Limitation of GC-MS | Page No. 70 |
| | Chapter – 6 Ion exchange chromatography | |
| 6.1 | Introduction | Page No. 71 |
| 6.2 | Matrix used | Page No. 71 |
| 6.2.1 | Cation exchanger | Page No. 72 |
| 6.2.2 | Anion exchanger | Page No. 73 |
| 6.3 | Principle | Page No. 74 |
| 6.4 | Steps in an ion exchange separation | Page No. 75 |
| 6.5 | Applications | Page No. 77 |
| 6.5.1 | Separation of proteins using cation exchanger | Page No. 78 |
| 6.5.2 | Softing of hard water | Page No. 79 |
| | References | Page No. 80 |

# Chapter 1 - Chromatography
## Basic concepts and its application

## 1.0 General Introduction

The most general method for removal of interference involves its physical separation from the analyte. Well-known methods for separations include distillation, crystallization, sublimation, solvent extraction, and chemical or electrolytic precipitation. However, the most widely used to eliminate interferences is chromatography. Chromatography was invented and named by the Russian botanist M. Tswett. He used the techniques to separate various plant pigments such as chlorophylls and xanthophyll by passing a solution of these compounds through a glass column packed with finely divided calcium carbonate. The separated species appeared as coloured bands on the column, which accounts for the name he chose for the method. Chromatographic methods have a wide importance that allow the separation, identification and determination of closely related components of mixtures. Chromatographic separations are relatively simple, fast and be completed in a few minutes. It give an accurate and reproducible result so that they are also useful for method development and impurity profiling of various synthesized compounds.

## 1.1 Principle of Chromatography

"Chromatography is a physical method of separation in which components in a mixture is separated based on different migration rates or different partition coefficient that can be possible by using two phases one is stationary phase and second is mobile phase".

In all chromatographic separation, the sample is dissolved in a solvent called mobile phase which may be gas, liquid and super critical fluid. This mobile phase is then forced through an immiscible phase which is fixed in a column called stationary phase. Between this two phases the components of the sample separated themselves. Those components strongly retained by stationary phase (their affinities towards the stationary phase is strong) move slowly (retention time is high), it is less eluted while other components that are weakly held by stationary phase travel rapidly (retention time is less), it is eluted first.

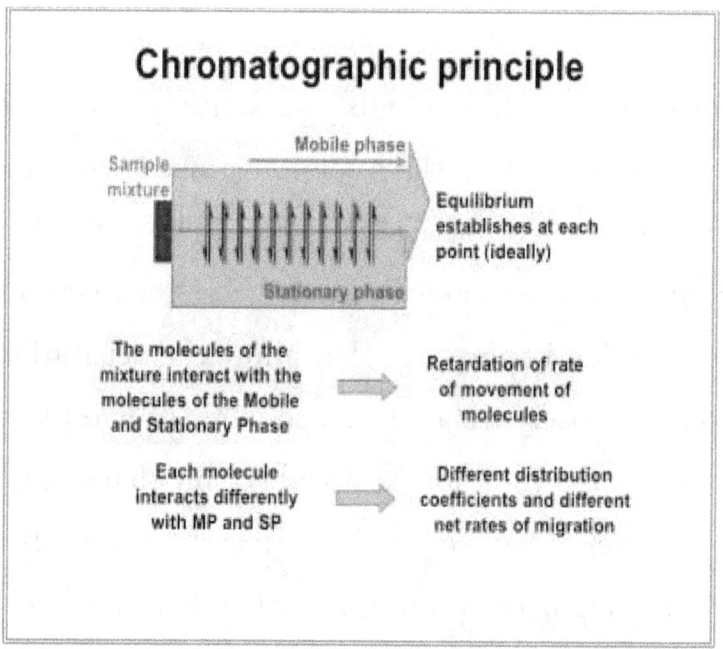

*Figure 1.1 Chromatographic Principle*

## 1.2 Classification of Chromatography

Common classification of chromatographic methods according to the nature of the stationary phase and mobile phase. It is also according to the mechanism of separation (principal) of the chromatography.

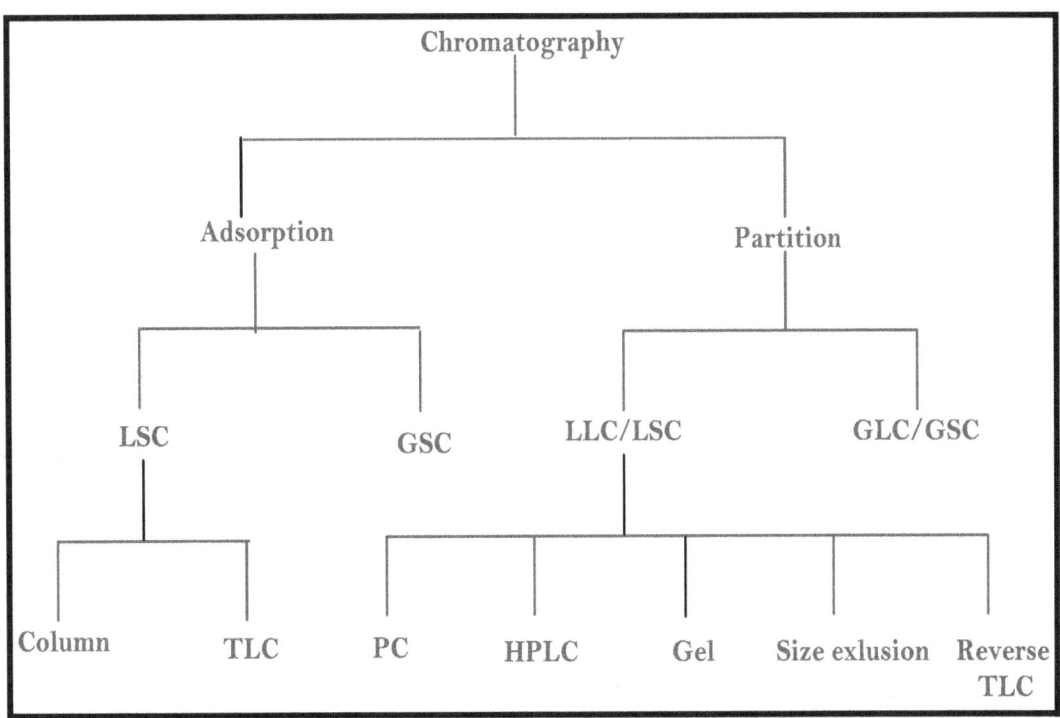

*Figure 1.2 Classification of Chromatographic techniques*

## 1.3 Distribution Coefficient (K)

All chromatographic separations are based upon differences in the extent to which solutes are partitioned between the mobile and stationary phase. The equilibria involved can be described quantitatively by means of a temperature dependent constant, the **partition or distribution coefficient (K)**

$$K = C_s / C_M \quad \ldots\ldots\ldots\ldots\ldots\ldots\ldots\ldots\ldots(1)$$

Where Cs is the analytical concentration of a solute in the stationary phase and $C_M$ is its concentration in the mobile phase. In the ideal case, the partition ratio is constant over a wide range of solute concentrations that is, Cs is directly proportional to $C_M$.

If K is less than 1 then the separation peak of the solute would be below the centre of the column (migration rate of the solute molecule is higher). If K is greater than 1 the separation peak of the solute would be the centre of the column (migration rate of the solute molecule is lower as compared to eluted peak). It is noted that K is affected by changing the distribution of components between the phases so that separation of components is also affected. By appropriate change in stationary phase, mobile phase or both, K can be also altered.

## 1.4 Chromatographic Terms

**1.4.1 Chromatogram:** It is an electronic visualization of chromatograph or a plot of the detector's signal as function of elution time or volume.

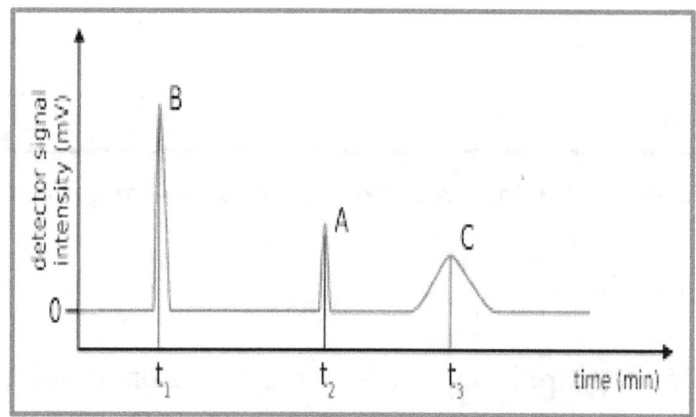

*Figure 1.3 Typical chromatogram of detector response as a function of retention time*

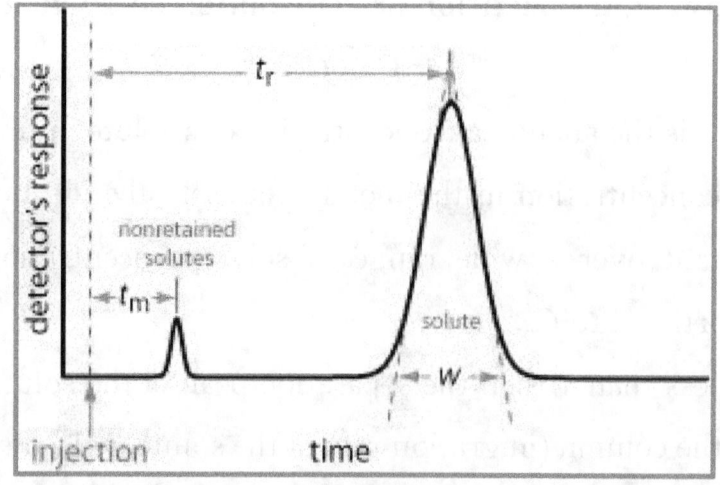

*Figure 1.4 Illustration of chromatogram with void time ($t_m$), retention time ($t_r$) and baseline width (w) for a solute*

### 1.4.2 Retention Time ($t_R$):
It is a time taken by a solute molecules to move from the point of injection to the detector. The retention time is assigned to the corresponding solute peak. The retention time is a measure of the amount of time a solute spends in a column.

### 1.4.3 Retention volume ($V_R$):
The retention time also can be measured indirectly as the volume of mobile phase eluting between the solute's introduction and the appearance of the solute's peak maximum. This is known as retention volumn. The volume of mobile phase required to move the solute molecules from the point of injection to the detector. Dividing the retention volume by mobile phase's flow rate (u) gives retention time.

$$\text{Retention time} = \text{retention volume} / \text{flow rate of mobile phase.}$$

### 1.4.4 Baseline width (w):
The width of a chromatographic peak of solute molecules measured at the baseline (w). It is measured by the intersection with the baseline of tangent line drawn through the inflection points of either side of chromatographic peak. Baseline width is measured in units of retention time or volume. Beside the solute peak, small peak eluted soon after the sample injection into mobile phase. This small peaks moves through the column at the same rate as mobile phase because they do not interact with the stationary phase, they are considered as unretained molecules. The retained molecules spends a time ($t_s$) in the stationary phase.

### 1.4.5 Void time or void volume ($t_M$):
It is a time or volume required for unretained molecules to move from the point of injection to detector is called void time or void volume.

### 1.4.6 Chromatographic resolution ($R_s$):
The goal of chromatography is to separate a sample into a series of chromatographic peaks, each representing a single component of the sample. Resolution is quantitative measurement of degree of separation between two chromatographic peaks, A and B, is defined as

$$Rs = \frac{t_{r,B} - t_{r,A}}{0.5\,(w_B - w_A)} = \frac{2\Delta t_r}{w_A - w_B} \quad \ldots\ldots\ldots\ldots\ldots (2)$$

As Shown in Figure 5, the degree of separation of peaks with an increase in $R_s$,

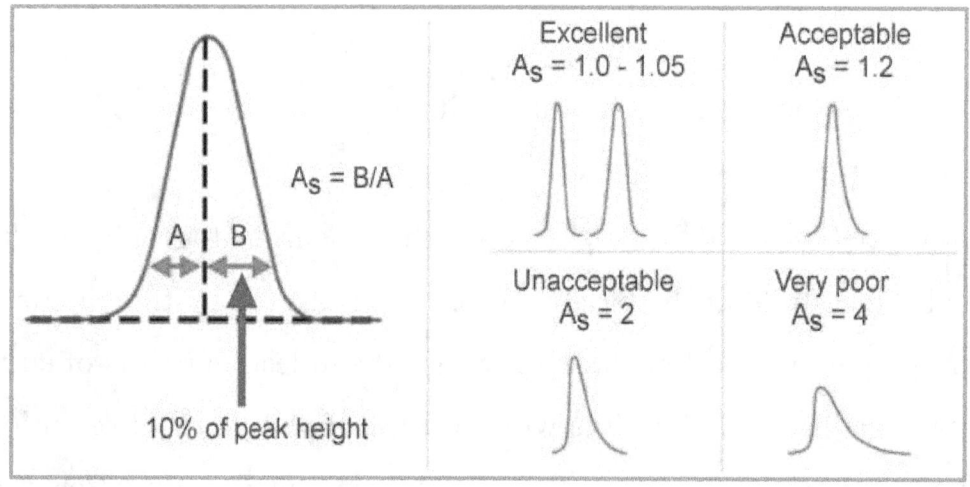

**Figure 1.5 Illustration of three chromatographic peak resolution**

From equation 2, it is clear that resolution may be improved either by increasing $\Delta t_R$ or by decreasing $w_A$ or $w_B$. Increases $\Delta t_R$ by enhancing the interaction of the solute with column or by increasing the column's selectivity for one of the solutes. Peak width is kinetic effect associated with solute's movement within and between the mobile phase and stationary phase. The effect is governed by several factors that are collectively called column efficiency.

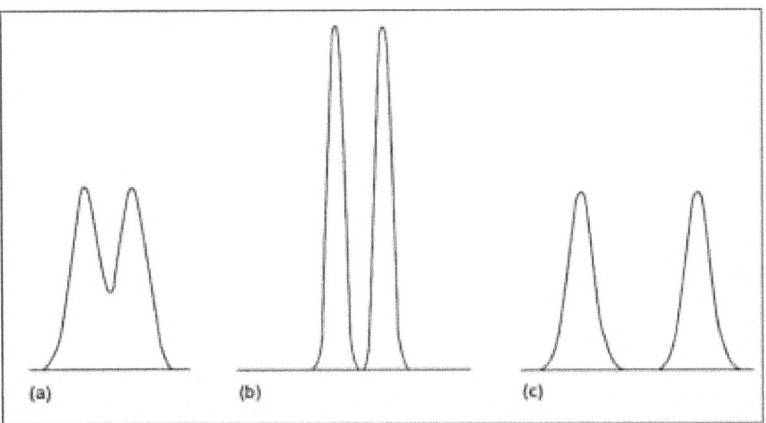

***Figure 1.6 Illustration of improving chromatographic resolution (a) Poorly resolved solutes (b) Improvement in resolution due to increases in column efficiency (c) Improvement in resolution due to change in column selectivity***

## 1.4.7 Retention factor (k) or capacity factor

The retention factor (k) measure of how strongly a solute is retained on the column. A solute's retention factor can be determined by measuring the void time of unretained molecule and retention time of retained molecule. The average linear migration rate (v) of the solute molecule through the column can be calculated by

$$v = \frac{L}{t_r} \quad \ldots\ldots\ldots\ldots\ldots(3)$$

Similarly average linear rate or velocity (μ) of the mobile phase can be calculated by

$$\mu = \frac{L}{t_M} \quad \ldots\ldots\ldots\ldots(4)$$

L = length of the column, $t_R$ = retention time of the solute molecule and $t_M$ = void time of the unretained solute molecule

Retention factor is the ratio of retention time of a solute spends on the column to the void time of an unretained solute. An unretained solute molecules has no affinity towards a stationary phase so that elutes with the solvent front at a time ($t_M$) which is known as "dead time" or "hold up time". Retention factor is independent of some of variable factors including small flow rate variables and column dimensions. Because of this useful parameter when comparing retention of chromatographic peak obtained using different columns. Retention factor (k) can be given equation

$$k = \frac{t_R - t_M}{t_M} = \frac{t_R'}{t_M} \quad \ldots\ldots\ldots\ldots\ldots(5)$$

$t_R$ = retention time, $t'_R$ = adjusted retention time and $t_M$ = retention time of unretained compound

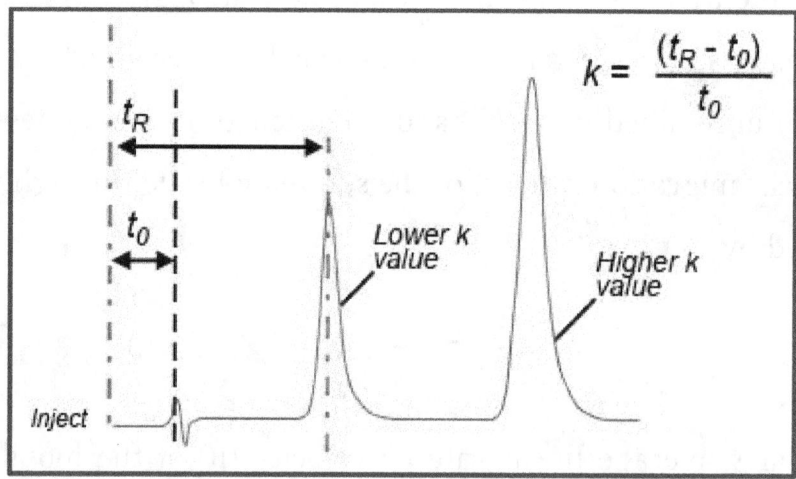

**Figure 1.7 Illustration of chromatogram with retention factors (k)**

Retention factor (k) is equal to zero for the unretained solute. Retention factor (k) is lower for the solute molecule which has less affinity towards stationary phase while k is higher for the solute which has high affinity towards stationary phase.

### 1.4.8 Column Selectivity (α)

The relative selectivity of a chromatographic column foe a pair of solutes is given by selectivity factor, which is define as

$$\alpha = \frac{k_1}{k_2} = \frac{t_{R2}-t_0}{t_{R1}-t_0} \quad \ldots \ldots \ldots \ldots \ldots (6)$$

The selectivity factor is a measure of the time or distance between the maxima of two peaks. It is usually measured as a ratio of the retention factors of the two peaks. Selectivity is always greater than one. When it is equal to one than the peaks are co-eluting. The greater selectivity value, separation between two peaks are higher.

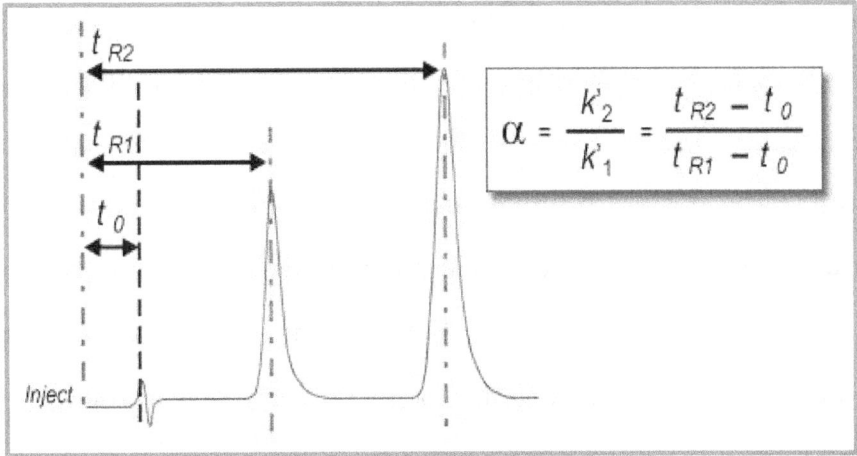

**Figure 1.8 Illustration of chromatogram with selectivity (α)**

Selectivity can be affected by changing the different solvents, stationary phase and temperature.

### 1.5 Plate theory (efficiency of column):

The efficiency of the column can be explained by plate theory. According to the plate theory developed by Martin and Synge, a chromatographic column consists a number of theoretical plates and equilibration of the solute between stationary and mobile phase take place at each of these plates. The movement of the solute down the column is assumed by a series of stepwise transfers between one plates to other.

The efficiency of separation of the column is increased as number of theoretical plate increases. This is because the number of equilibrations will also increases means better quality of separation. Efficiency of the column is depends on no of theoretical plates (N) and plate height (H) or HETP (Height Equivalent to a theoretical Plate) given by

$$N = \frac{L}{H} \quad \quad \quad (7)$$

The efficiency of the column increases as number of theoretical plates (N) becomes greater and as the plate height (H) becomes smaller.

The plate theory successfully accounts for the Gaussian shape (due to dispersion of the peaks) of chromatographic peak and their relative movement down a column. The breadth (width) of Gaussian curve is described by the standard deviation ($\sigma$) or its variance ($\sigma^2$). So it is convenient to define efficiency of the column in terms of variance per unit length of the column. That is, the plate is given by

$$H = \frac{\sigma^2}{L} \quad \quad \quad (8)$$

The theory was ultimately abandoned in favour the rate theory because its fails to accounts the peak broadening. Put the equation (8) in equation (7)

$$N = \frac{L^2}{\sigma^2} \quad \quad \quad (9)$$

## 1.5.1 Calculation of Chromatographic peak width

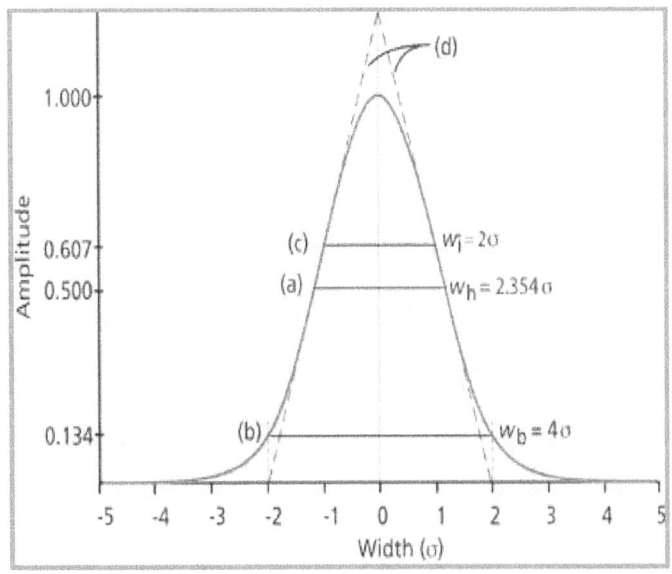

*Figure 1.9 Illustration of Normal Gaussian curve characteristic*

The breadth (width) of Gaussian curve is described with units of length and time. If breadth of curve is described with units of length then σ sign is used while for time, τ sign is used. The relation between this two standard deviations are related by

$$\tau = \frac{\sigma}{L/tr} \quad \quad \quad \ldots\ldots\ldots\ldots\ldots (10)$$

The width of chromatographic peak is W=4 τ.

$$\frac{W}{4} = \frac{\sigma}{L/tr} \quad \quad \quad \ldots\ldots\ldots\ldots\ldots (11)$$

Rearrange the equation (11)

$$\sigma = \frac{LW}{4tr} \quad \quad \quad \ldots\ldots\ldots\ldots\ldots (12)$$

Making a square of equation (12) and substitute the value of (12) into equation (9)

$$N = 16 \left(\frac{t_r}{w}\right)^2 \quad \quad \quad \ldots\ldots\ldots\ldots\ldots (13)$$

Alternatively, the number of theoretical plates can be approximated as

$$N = 5.454 \left(\frac{t_r}{w_{1/2}}\right)^2 \quad \dots\dots\dots\dots\dots (14)$$

## 1.6 Rate Theory (Band Broadening)

Band Broadening of chromatographic peak can be explained by rate theory. It describes the more realistic description of what actual processes work inside a column. We know that the chromatographic peaks look like Gaussian or normal error curves but some peaks are non-ideal and exhibit tailing and fronting.

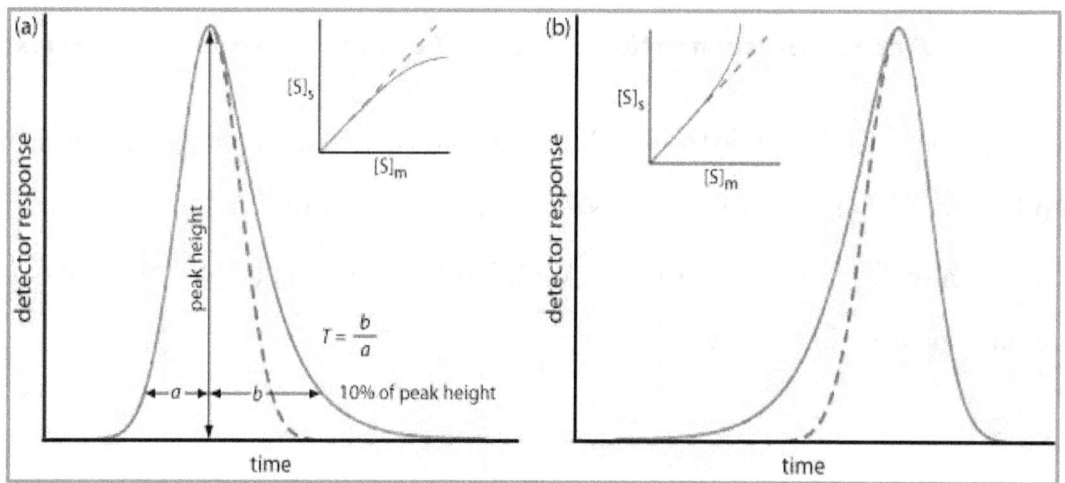

**Figure 1.10 Illustration of effect of telling and fronting on chromatogram**

Band broadening reflects a loss of column efficiency. Band broadening takes due to the slower the mass transfer processes occurs while a solute migrates through a column. (Time taken for the solute to equilibrate between the stationary and mobile phase). The resulting band shape of a chromatographic peak is affected by the rate of elution.

If we consider the various mechanisms which contribute to band broadening, we arrive at the Van Deemter equation for plate height;

$$\text{HETP} = A + B/u + C u \quad \ldots\ldots\ldots\ldots\ldots (15)$$

Where $u$ is the average velocity of the mobile phase. *A*, *B*, and *C* are factors which contribute to band broadening.

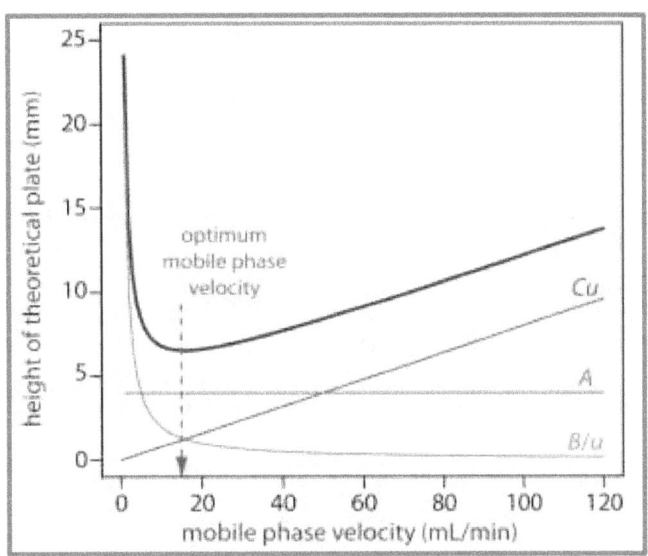

*Figure 1.11 Plot of the height of a theoretical plates as a function of mobile phase velocity using van Deemter equation. The contributions to the terms A, B/u and Cu also are shown*

### 1.6.1 Eddy Diffusion (Multipath diffusion)

Solute molecules will take different paths and travel through the stationary phase to reach at end of the column. This multiple path effect is called eddy diffusion which would be independent of solvent velocity.

### 1.6.2 Longitudinal diffusion

The concentration of solute molecules is lower at the edges of the band than at the centre. Solute molecules diffuses out from the centre to the edges.

This causes band broadening. If the velocity of the mobile phase is high then the analyte spends less time on the column, which decreases the effects of longitudinal diffusion.

### 1.6.3 Mass transfer

A chromatographic separation occurs solutes move between stationary phase and mobile phase. For the movement of the solute from one phase to other phase, it must diffuse first it is called **mass transfer.** The solute molecule takes a certain amount of time to equilibrate between the stationary and mobile phase.

The band broadening occurs whenever movement of solute molecules is not fast enough to maintain the equilibration. If the velocity of the mobile phase is high, and the analyte has a strong affinity for the stationary phase, then the analyte in the mobile phase will move ahead of the analyte in the stationary phase. The band of analyte is broadened. The higher the velocity of mobile phase, the worse the broadening becomes.

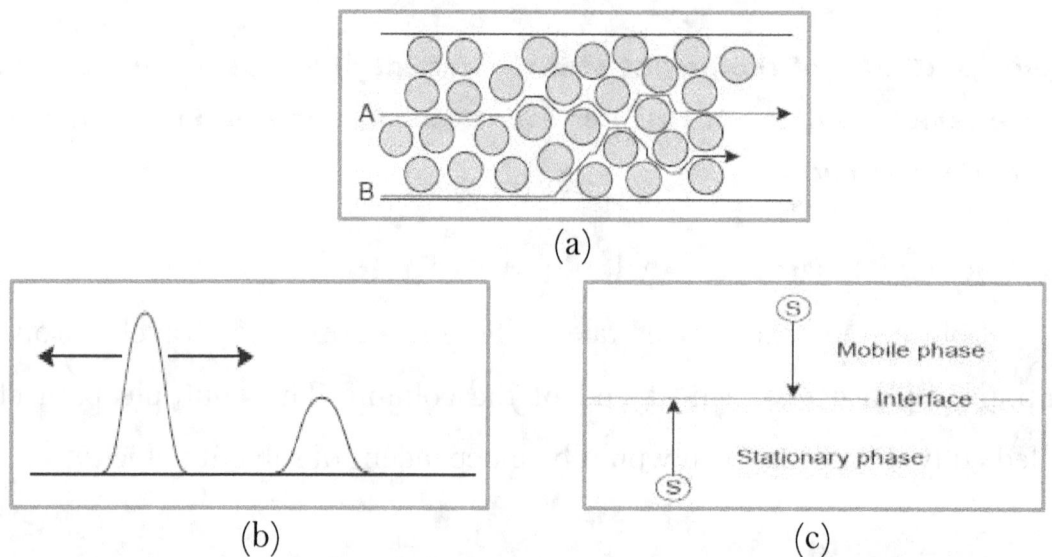

*Figure 1.12 Illustration of band broadening due to (a) multiple paths (b) longitudinal diffusion (c) mass transfer*

# Chapter 2 - High Performance Liquid Chromatography (HPLC)

## 2.1 Introduction

HPLC is a chromatographic technique that can separate a mixture of compounds under high pressure. It is used to identify, quantify and purify the individual components of a mixture.

## 2.2 Why it is superior to traditional column chromatography?

Traditional column chromatography has slow flow rate of mobile phase and separation of the components under the gravity take a long time. It is not useful separation of multiple components in sample. This problem is overcome by using a applying a high pressure to column. The principle advantage of the HPLC is the speed at which separation take place. Because of the decreases in time, diffusion in the column is reduced and resolution improved.

Column packing, limited length of column and high pressure makes HPLC superior than other. The smaller size of particles (3-5 μ) which decreases the eddy diffusion (multiple path) in the Van Deemter relationship shows the better resolution. Limited length of column is use to overcome the problem of peak broadening. In HPLC, pressure used normally 300-600 bar. But highly pressure (faster flow of mobile phase) is not possible because its creates a back pressure which is sufficient to damage the column matrix. So it need to control the pressure by using pumps. The pressure depends on the column used. The pressure varied to provide the optimum linear flow rate of the mobile phase and smallest theoretical plate height. So that resolving power of column increased.

It is more versatile than GC because it is not limited to volatile (mostly organic solvents) and thermally stable compounds and choice of the mobile phase and stationary phase is wider.

## 2.3 Principle

**It is similar to the adsorption chromatography (LSC).** It is based on the interactions between the solute and fixed active site on finely divided solid absorbent used as stationary phase.

**But widely used type of HPLC is partition chromatography (LLC)** in which stationary phase is a second liquid that is immiscible with the liquid mobile phase. In the past, most of application of LC was only for non-ionic, polar compounds having low molecular weight because the use of liquid-liquid columns. But this problem have been replaced by using a liquid-bonded-phase columns which is nowadays used in HPLC.

## 2.4 Column packing (stationary Phase)

Three type of packing is used in HPLC.

### Pellicular

The pellicular particles are spherical, nonporous, glass or polymer beds with typical diameters of 30 to 40 µm. But Pellicular particles has been replaced by small **porous micro particles** having diameters ranging from 3 to 10 µm. These particles are composed of silica, alumina, synthetic resin or ion exchange resin. But porous micro particles columns having a high efficiency but low sample capacity. Both pellicular and porous micro particles are used a solid support for the liquid. **But disadvantage** of this solid support is that the developing solvent may gradually wash off the liquid phase with repeated use. To overcome this problem liquid bonded phase have been developed.

### Liquid-bonded-Phase:

In liquid bonded phase, the liquid (long chain of hydrocarbon) is attached to solid support (silica or silicon polymer) by chemical bonding (covalent).

## 2.5 Bonded phase column (Silanization)

The particles (silica or silica based compositions) having diameters ranging 1.5 to 10 micron. 3 and 5 micron size particles are most common used. The surface of fully hydrolysed silica is made up of chemically reactive group silanol (Si-OH) group. This Si-OH gives undesirable polarity to the surface of silica which may lead to tailing of chromatographic peaks particularly for basic solutes. So there is need to coat the surface of silica by silanization.

In silanization, silixanes are formed by reaction of Si-OH groups on surface of silica with an organochlorosilane. Example such as

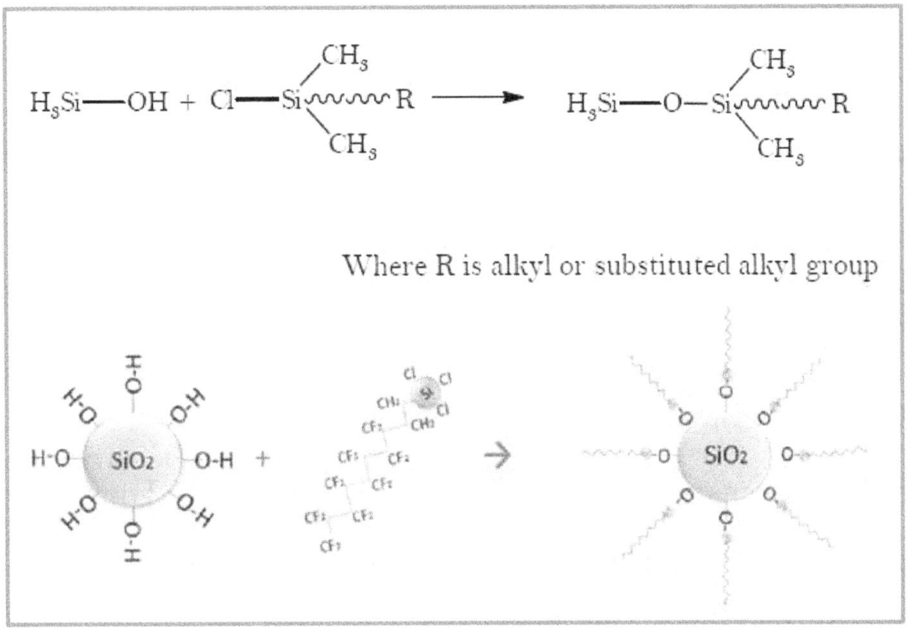

$$H_3Si-OH + Cl-Si(CH_3)(CH_3)-R \longrightarrow H_3Si-O-Si(CH_3)(CH_3)-R$$

Where R is alkyl or substituted alkyl group

*Figure 2.1. Illustration of Silanization*

The properties of a stationary phase are determined by the nature of the organosilane's alkyl group.

If R is a polar functional group, then the stationary phase will be polar. R contains a cyano ($-C_2H_4CN$), diol ($-C_3H_6OCH_2CHOHCH_2OH$), or amino ($-C_3H_6NH_2$) functional group. Since the stationary phase is polar, the mobile phase is a nonpolar or moderately polar solvent then the combination of a polar stationary phase and a nonpolar mobile phase is called **normal-phase chromatography.**

If R is a non-polar functional group, then the stationary phase will be non-polar. R contains *n*-octyl ($C_8$) or *n*-octyldecyl ($C_{18}$) hydrocarbon chain. Since the stationary phase is nonpolar, the mobile phase is a polar solvent then the combination of a non-polar solvent and polar mobile phase is called **reverse phase chromatography**. Most reverse phase separations are carried out using a buffered aqueous solution as a polar mobile phase. Because the silica substrate is subject to hydrolysis in basic solutions, the pH of the mobile phase must be less than 7.5.

Silanization is limited or less on surface of silica because of steric effects. The unreacted Si-OH groups lead tailing. The problem is overcome by further reaction with chlorotrimethylsilane. Silanization makes a column more retentive.

## 2.6 Types of HPLC or Separation modes

The four major separation mode that are used. They are distinguished based on the relative polarities of the mobile phase and stationary phases.

1. Normal Phase Chromatography
2. Reverse Phase Chromatography

## 2.6.1 Normal phase chromatography

In these mode, the column packing is polar (e.g. silica gel, cynopropyl, amino etc.) and mobile phase is non-polar (hexane, iso-octane, methylene chloride, ethyl acetate). The least polar (non-polar) compounds is eluted first, increasing the polarity of the mobile phase, the elution time is decreased. This techniques is useful for water sensitive compounds, geometric isomers, cis-trans isomers, class separations and chiral compounds.

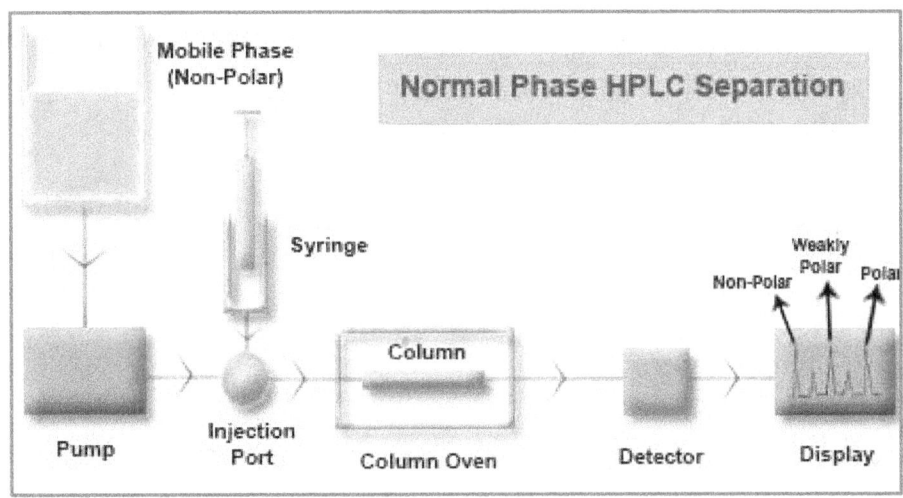

*Figure 2.2 Schematic diagram of normal phase chromatography*

## 2.6.2 Reverse phase chromatography

In these mode, the column packing is non-polar (e.g. $C_{18}$, $C_8$, $C_3$, phenyl, etc.) and mobile phase is polar (water, methanol, acetonitrile, THF and buffers). The most polar compounds elutes first, increasing mobile phase polarity, the elution time is increased. It is a most popular mode over 90% of chromatographic development is done by this mode. It can be used for non-polar, polar ionisable and ionic molecules. The major **advantage of revers phase** is that the water is used as mobile phase. Water is inexpensive, nontoxic, UV-transparent solvent compatible with biological solutes. Also mass transfer is rapid with non-polar stationary phases.

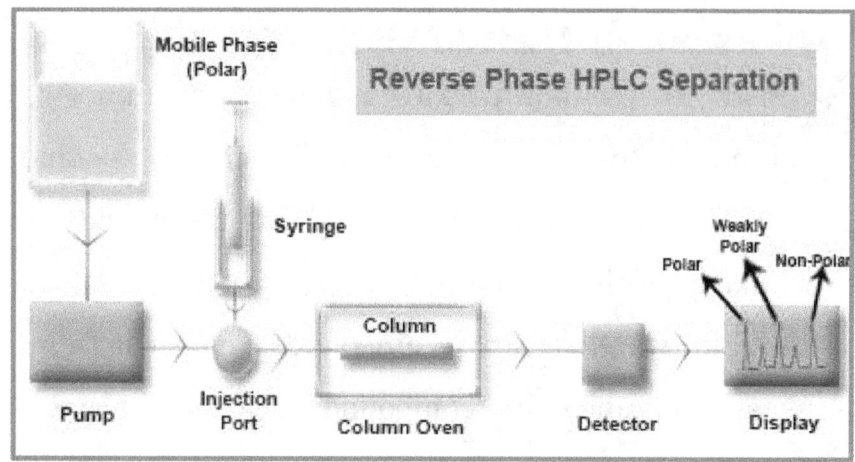

*Figure 2.3 Schematic diagram of reverse phase chromatography*

## 2.7 Elution modes

Generally separation of simple and mixture of components can be done with isocratic and gradient elution.

## 2.7.1 Isocratic elution

This techniques is generally used for simple separation of all organic and pharmaceutical products. Most wide used in a quality control applications to check out the purity of compounds. In this techniques the **mobile phase composition remains constant** throughout separation process. Chromatographic peak width increase with retention time (late eluting peaks get very flat and broad) which leads to the disadvantage.

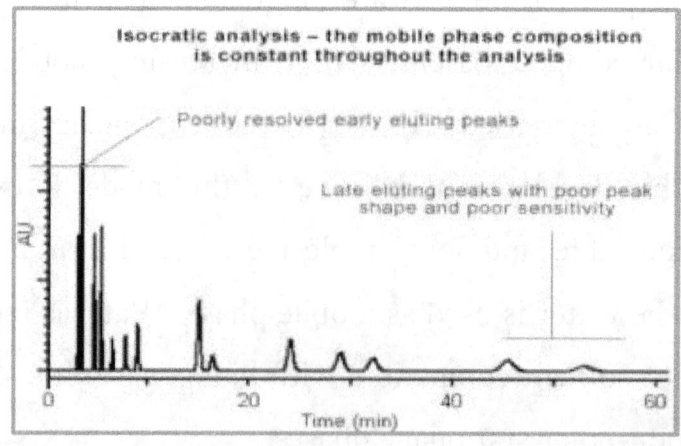

*Figure 2.3 Illustration of isocratic separation*

## 2.7.2 Gradient elution

This techniques is used for the separation of samples containing analytes with wide range of components (multi components). In such condition a constant mobile phase composition during analysis may not be provide an acceptable separation. To overcome these problems, the mobile phase composition is changed during analysis. One begins with a predominantly water-based mobile phase and then adds organic solvents as a function of time. Organic solvent increases the mobile phase strength and elutes components that are strongly retained on the reverse phase packing.

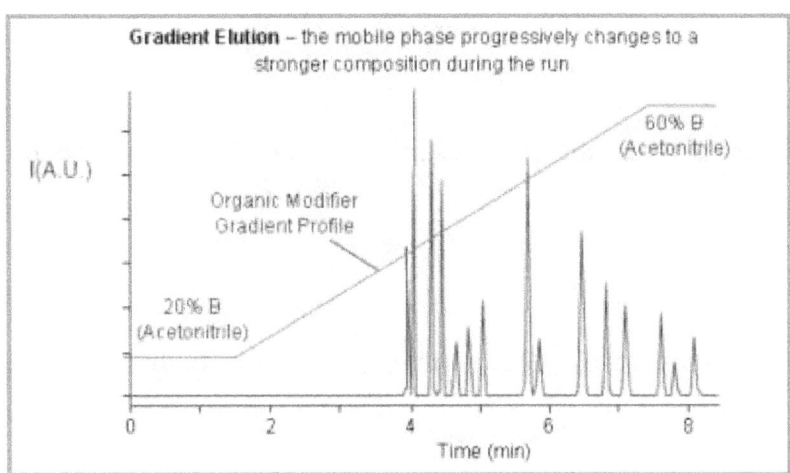

*Figure 2.4 Illustration of gradient separation*

## 2.8 HPLC Instrumentation

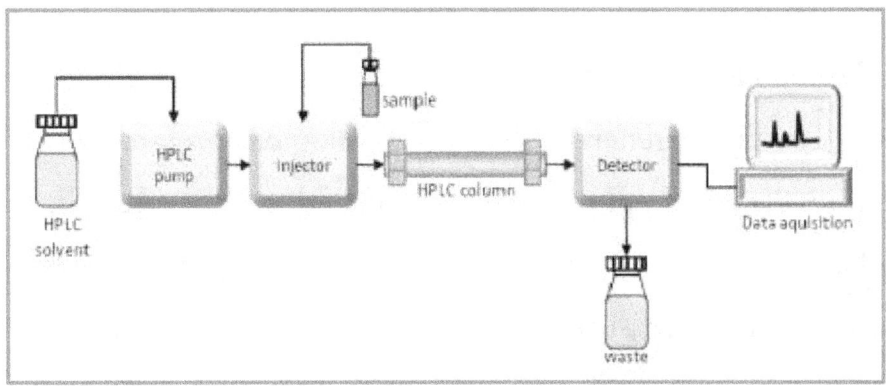

*Figure 2.5 Schematic diagram of HPLC instrument*

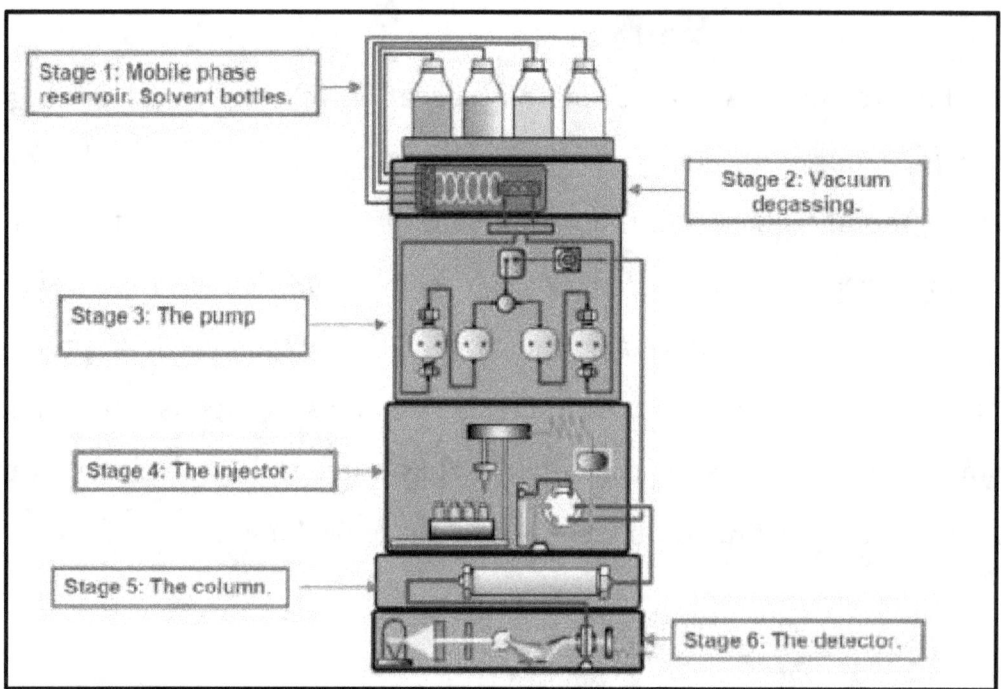

*Figure 2.6 Schematic diagram of HPLC instrument components*

## 2.8.1 Working Function

**Mobile Phase reservoir -** The appropriate solvents are stored in reservoir.

**Vacuum Degassing** – Morden HPLC system have online vacuum degassing to remove dissolved gas from the mobile phase. This gives less noisy baseline and better chromatographic performance.

**Pump** – Mobile phase from reservoir is allowed to enter the mixing chamber through pump where mobile phase components can be mixed in different ratio to alter the mobile phase characteristics which help to improve separation and resolution. The pump generally creates 300-600 bar pressure.

**Sample injector** – Sample is injected through a port into the high pressure mobile phase flow (1 to 2 ml/min), without interrupting mobile phase flow or reducing the system pressure.

**Column** – The separation of components takes place on the column which vary from 50-100 cm length and 1-2 internal diameter. The sample is partitioned between the mobile and stationary phase. The degree of partition between phases will govern its retention.

**Detector** – The detector measure some physico-chemical property (UV absorption) of mobile phase elutes from the column. The response of detector will changes as the sample components begin to eluate.

## 2.9 Choice of solvent or mobile phase preparation

### 2.9.1 Solvent miscibility

In reverse phase mode, different mobile phase composition is used. If immiscible solvents are used, mixing can become non-uniform and formation of an accurate and reproducible mobile phase composition is impossible. Immiscible solvents may cause de-wetting and leaves the traces which coat the internal components of the HPLC and causing a blockages. Immiscibility of solvent also affect capacity of pumping system. The pump produces inaccurate mobile phase composition resulting an unstable back pressure. Unstable back pressure drift the retention time and unstable baseline. To solve this problem, it is good practice to flush the system through with water containing increasing amount of isopropanol. Isopropanol is miscible with all common solvents both organic and aqueous.

### 2.9.2 Viscosity

Lower viscosity of solvent give narrow peaks due to improved mass transfer of analyte in the mobile phase. Viscosity is also important when considering system backpressure. The more viscous the solvent the higher the system back pressure.

### 2.9.3 UV cut-off

UV absorbance of an analyte is plotted against time. The higher the UV absorbance the higher analyte concentration at that time. In HPLC generally UV transparent solvent is used. (Solvent has no UV absorption at the wavelength at which measurement are to be taken.). Using a solvent with high UV cut-off at the selected wavelength for measurement can result in increased noise level and a loss in sensitivity. So that UV cut-off must be considered. A solvent with a UV cut-off below 200 nm is used for the measurement.

### 2.9.4 Polarity

Polarity index P' (quantitative measurement of solvent's polarity) is most useful for selection of mobile phase. Larger value of polarity index correspond to more polar solvents. Mobile phases of intermediate polarity can be fashioned by mixing together two or more of the mobile phases in Table 1.

**Table 2.1 Properties of HPLC Mobile phase**

| Properties of HPLC mobile phase | | |
|---|---|---|
| **Mobile Phase** | **Polarity Index (P')** | **UV cut off (nm)** |
| Cyclohexane | 0.04 | 210 |
| n-Hexane | 0.1 | 210 |
| Toluene | 2.4 | 286 |
| Diethyl ether | 2.4 | 218 |
| THF | 4.0 | 220 |
| Ethanol | 4.3 | 210 |
| Ethyl Acetate | 4.4 | 255 |
| Dioxane | 4.8 | 215 |
| Methanol | 5.1 | 210 |
| Acetonitrile | 5.8 | 190 |
| Water | 10.2 | - |

Polarity of binary solvents can be calculated by

$$P'_{AB} = \phi_A P'_A + \phi_B P'_B \quad \text{...................(2.1)}$$

Where $P'_A$ and $P'_B$ are the polarity indexes for solvents A and B, and $\phi_A$ and $\phi_B$ are the volume fractions of the two solvents.

> **Example: A reverse phase HPLC separation is carried out using a mobile phase mixture of 60% v/v water and 40% v/v methanol. What is the mobile phase's polarity index?**
>
> Solution:
>
> Polarity index of water is 10.2 and 5.1 for methanol. Using above equation 2.1, polarity index for a 60:40 water-methanol mixture is
>
> $$P'_{AB} = (0.60)(10.2) + (0.40)(5.1) = 8.2$$

## 2.9.5 Selectivity factor

To effect of a better separation between two solute it is often necessary to improve the selectivity factor ($\alpha$). Two approaches are commonly used for this improvement of selectivity factor.

## pH levels

When a solute is a weak acid or weak base, adjusting the pH of the aqueous mobile phase can lead to significant changes in the retention time of solute. Example of reverse-phase separation of *p*-aminobenzoic acid and *p*-hydroxybenzoic acid on a nonpolar C18 column.

At more acidic pH levels, both weak acids are present as neutral molecules. Because they partition favourably into the stationary phase, the retention times for the solutes are fairly long. When the pH is made more basic, the solutes, which are now present as their conjugate weak base anions, are less soluble in the stationary phase and elute more quickly. Similar effects can be achieved by taking advantage of metal–ligand complexation and other equilibrium reactions.

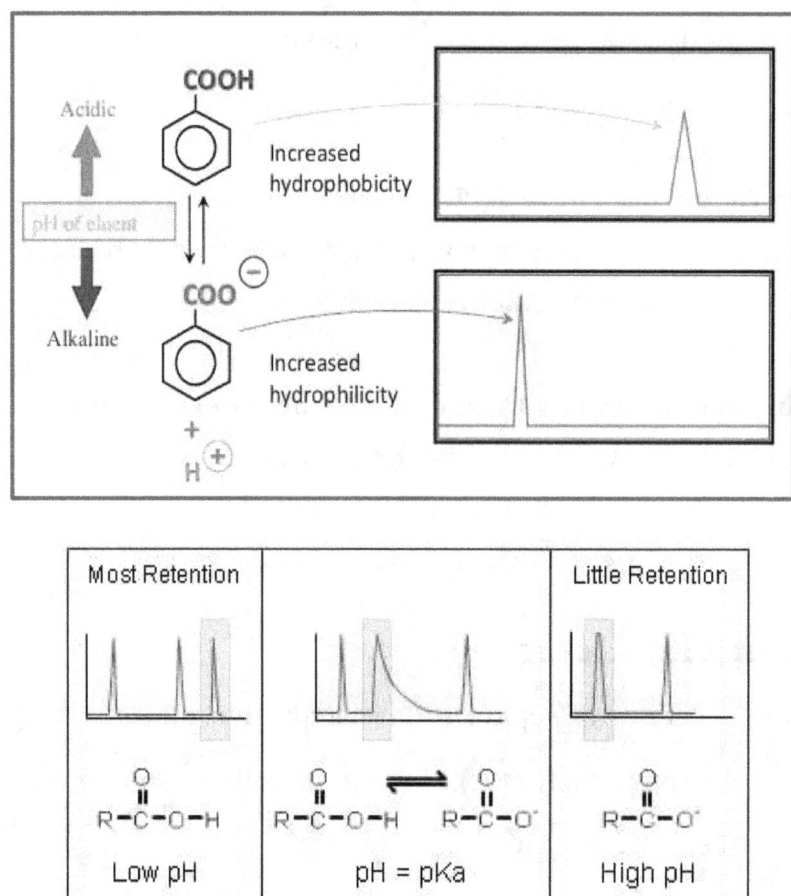

*Figure 2.7 pH of eluent and retention of ionic solute*

## Solvents Pair

In a reverse-phase separation, separation of solutes is accomplished by making a pair of solvent mixed with water which is used to adjusting retention times. Besides methanol, other common solvents for adjusting retention times are acetonitrile and tetrahydrofuran (THF). The properties of these three solvents can be mapped in a solvent selectivity diagram as shown.

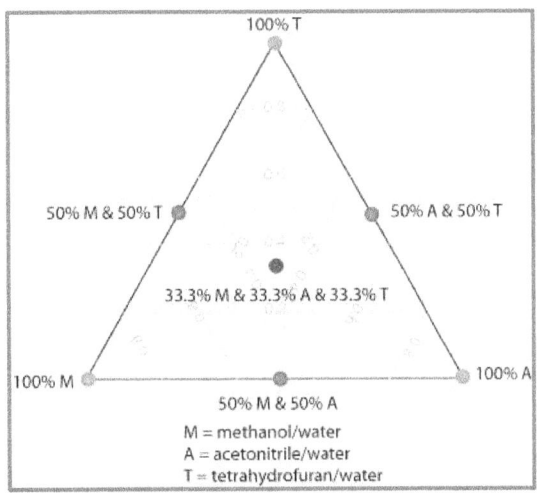

*Figure 2.8 Solvent triangle for optimizing RP- HPLC separation.*

These diagram shows that THF has greatest dipole interaction as followed by acetonitrile and then methanol. So that THF consider as strongest organic modifier and causing a largest changes in retention time. Methanol is most basic while acetonitrile is most acidic among three solvents. When changing the composition of organic modifier with water it is more convenient to use the information from a **"nomogram"**. A nomogram compares the relative strengths of solvents pairs. In this way an equivalent mobile phase composition can be used that has same eluting power as the original (overall separation will occur in the same frame) but with differing selectivity. In such causes the phases being compared are said to be "**isoeluotropic**".

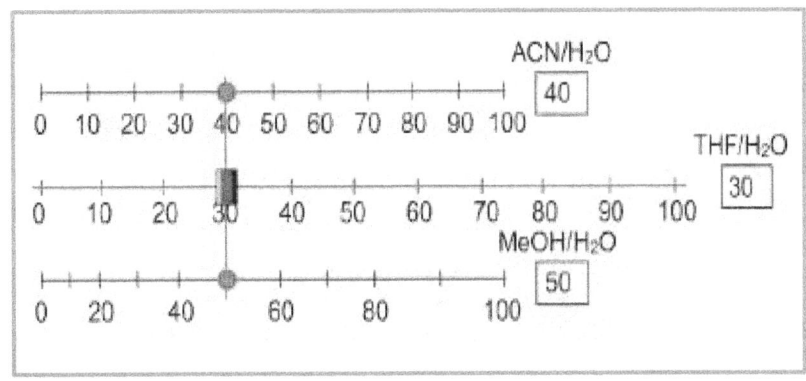

*Figure 2.9 Illustration of iso-eluogram (nanogram)*

## 2.10 Importance of mobile phase degassing

Mobile phase degassing is used to remove dissolved gases such as $N_2$ and $O_2$, and small particulate matter, such as dust from the mobile phase. By removing oxygen prevent the possible oxidative degradation of mobile phase and sample. Air in the mobile phase also cause a variety of problems. Air in the pumping system can cause a loss of pumping efficiency. It produces a back pressure which drifts the retention time, poor solvent mixing and noisy baseline appearance. Air with a high pressure will destroy column packing and causing a channels. This can lead to poor efficiency and fronting peaks.

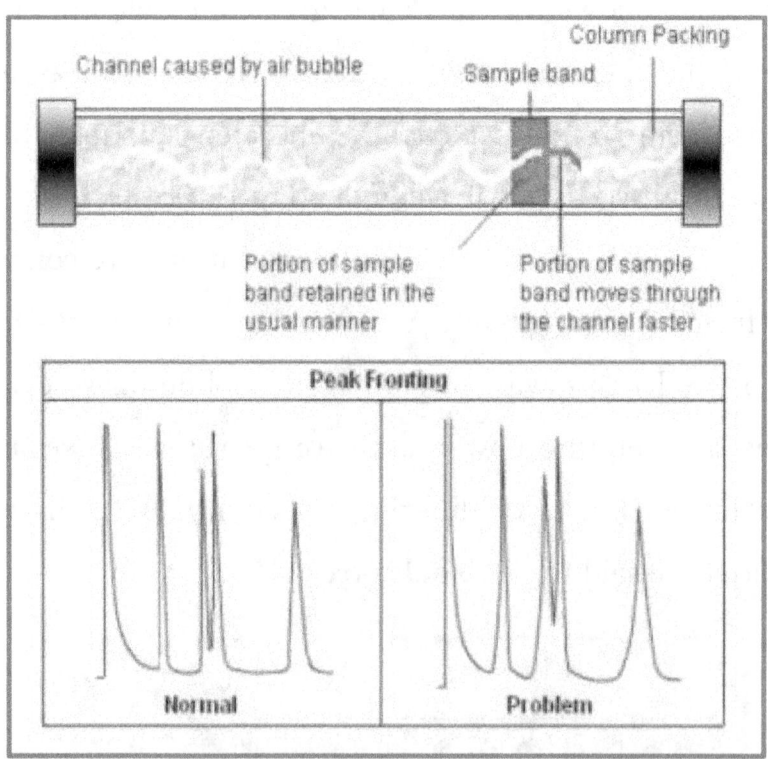

*Figure 2.10 Destroy of the column packing without using degasser*

There are four commonly used off-line methods for degassing: 1) Helium Sparging 2) Ultrasonic Degassing 3) Vacuum Degassing 4) Refluxing

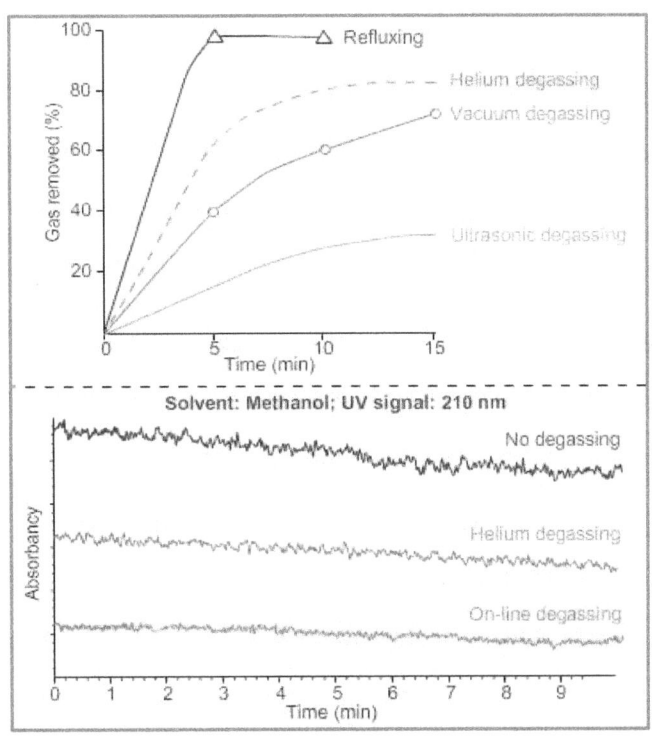

*Figure 2.11 Commonly degassing techniques used in HPLC*

## 2.11 HPLC Pumping

After the removal of the impurities and dissolved gases by degassing and filtering, mobile phase is pulled from their reservoirs to column by action of a pump. Most HPLC instruments use a **reciprocating pump.**

## Working function of reciprocating pump

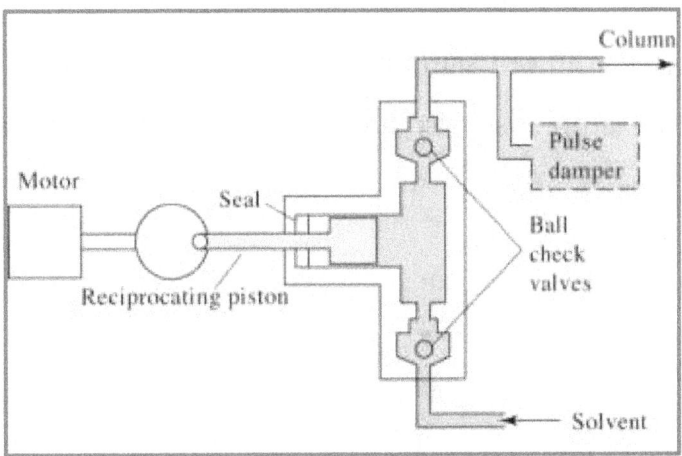

*Figure 2.12 Illustration of working function of reciprocating pump*

Reciprocating pump consisting of a piston whose back-and-forth movement is capable both of maintaining a constant flow rate of up to several millilitres per minute and producing a high output pressure needed to push the mobile phase through the chromatographic column. A solvent proportioning valve controls the mobile phase's composition, making possible the necessary change in the mobile phase's composition when using a gradient elution. The back and forth movement of a reciprocating pump results in a pulsed flow that contributes noise to the chromatogram. To eliminate this problem a pulse damper is placed at the outlet of the pump.

## 2.12 Sample Injection technique

In HPLC, operating pressure is sufficiently high. So it is impossible to inject the sample in the same manner as in gas chromatography. Instead of the sample is introduced using **a loop injector** means sample is loaded into a short section of tubing and injected into a column by directing the mobile phase through the loop. Sampling loops are interchangeable and available with volumes ranging from 0.5 µl to 2 ml. In the load position, the sampling loop is open and isolated from the mobile phase. A syringe with sample capacity is used to place the sample in to loop. Any extra volume of sample exits through the waste line. After loading of sample, injector is turn to the injection position. In this position, the sample is introduced into the flow of mobile phase and then passed through the column.

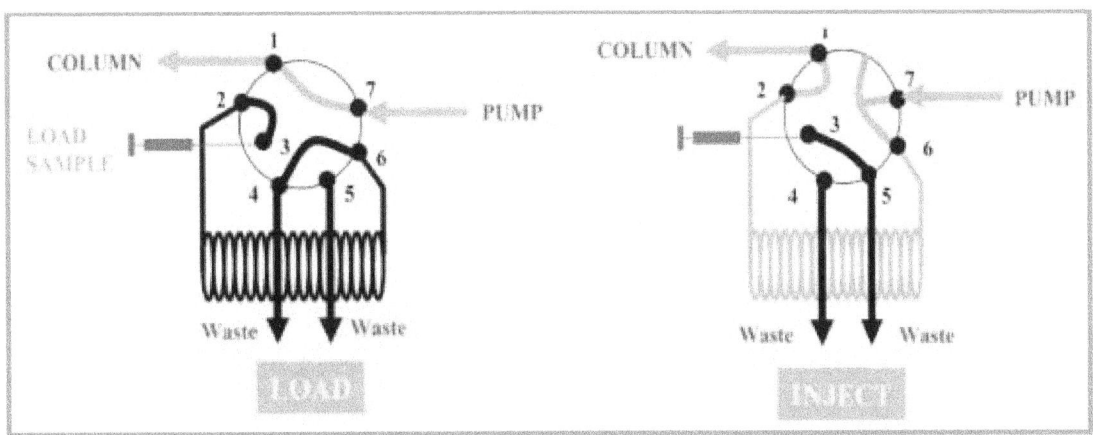

Figure 2.13 Schematic diagram of a loop injector in the
(a) Load and (b) Inject positions

## 2.13 HPLC column

An HPLC typically includes two columns: an **analytical column** responsible for the separation and a guard column. The **guard column** is placed before the analytical column, protecting it from contamination.

### Analytical column

The most commonly used columns for HPLC are constructed from stainless steel with internal diameters between 2.1 mm and 4.6 mm, and lengths ranging from approximately 30 mm to 300 mm. These columns are packed with 3–10 mm porous silica particles that may have an irregular or spherical shape. Typical column efficiencies are 40,000–60,000 theoretical plates/m.

### Guard column

Usually, a short column is placed before the analytical column to increases the life of analytical column by removing not only the particulate matter and contamination from the solvent but also sample components that binds with stationary phase. The packing of the guard column is similar to the packing of analytical column but the size of particles is usually larger to minimize the pressure drop.

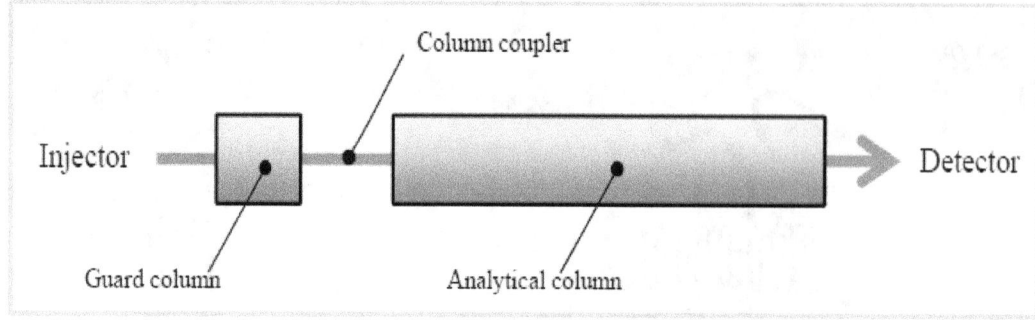

**Figure 2.14 Schematic diagram of coupling of guard and analytical column**

## 2.14 HPLC Detectors

The function of the detector is to detect the individual molecules that elute from the column and convert the data into an electrical signal. The detection process in LC has more complication than GC because GC has a specialized universal FID and TCD detector for the detection of the lowest amount of the solute while LC has no universal detector for the detection. Suitable detectors in HPLC can be broadly divided into following two classes. Bulk property and solute property.

Bulk property detector respond to a mobile phase properties such as refractive index dielectric constant or density that is modulated by a presence of solute so that have poor sensitivity and having a limited detection range.

Solute property detector respond to particular chemical or physical properties of solute such as UV-absorbance, fluorescence and diffusion current that cannot be affected by changes in mobile phase composition so that such type of detectors are ideally independent of the mobile phase. So that bulk property detectors like UV-visible absorbance, florescence, electrochemical and mass spectrometry.

### 2.14.1 UV-visible Detector

The most widely used, most versatile, having the best sensitivity and non-destructive. It is a relatively insensitive to changes of temperature and flow rate of mobile phase. These detector uses of deuterium and tungsten lamp as source to produce wavelength range of 190-800 nm. These detector having a simple designs, in which the analytical wavelength is selected using appropriate filters. These detector works by measuring the amount of light absorbed by a solute molecule passed through a small HPLC flow cell held in the radiation. The resulting chromatogram is a plot of absorbance as a function of elution time. Both single beam and double beam instruments are commercially available. The earliest detectors had a fixed detection wavelengths (254 nm or 280 nm) so they are restricted to solutes. But now a days this type of detectors have been replaced by scanning and diode array detectors.

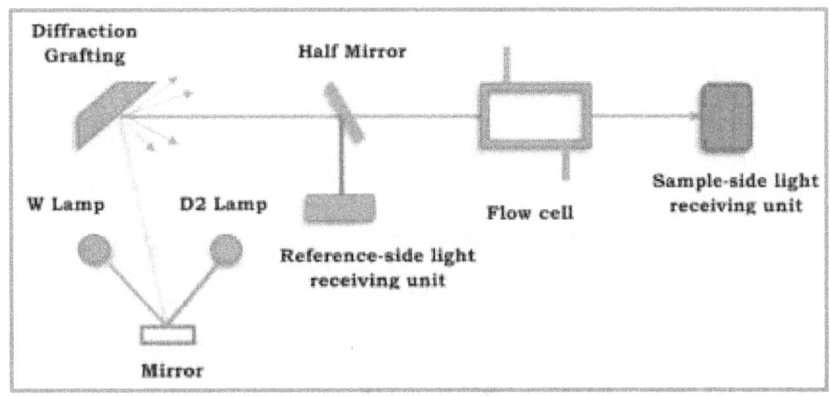

**Figure 2.14 Schematic diagram of a UV-Visible detector**

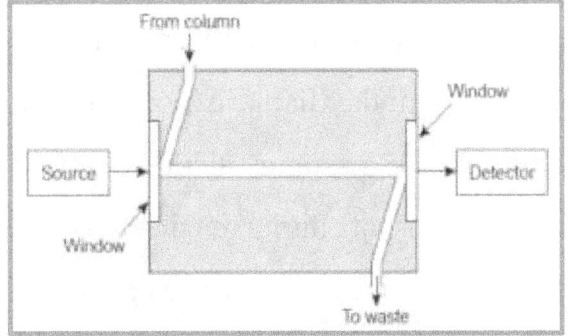

**Figure 2.15 Schematic diagram showing a flow cell**

## 2.14.2 Photo diode array (PDA) or Diode array detector

UV-Visible detector has only one sample side light receiving section while PDA has multiple photodiode arrays from sample side to obtain information over a wide range of wavelengths at one time. PDA differ from the UV-visible detector. In PDA, polychromatic light is passed through a HPLC flow cell and emerging radiation is diffracted by a grating to fall on an array which consist of a series of over a thousand semiconductor sensors. Each of this semiconductor unit receives a different wavelengths. This changes in wavelengths can be recorded as a separated compounds passes through a HPLC cell.

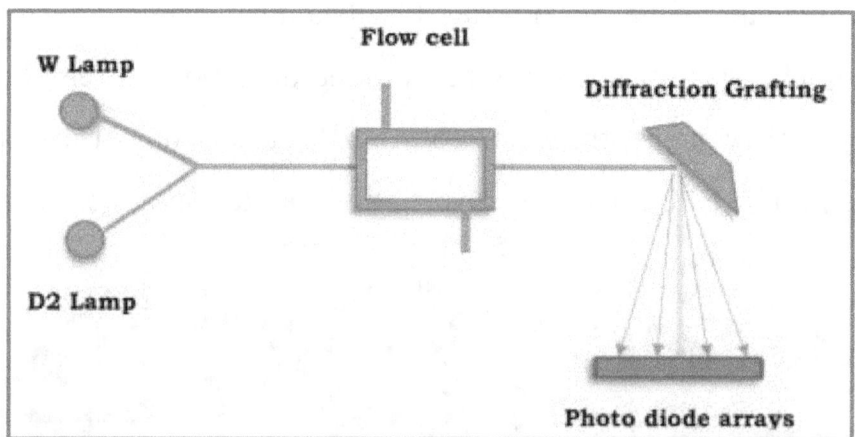

**Figure 2.16 Schematic diagram of a DAD optical system**

## 2.14.3 Refractive index detector

RI detector are based on changes of refractive index of the eluent from the column with respect to pure mobile phase. This detector is extremely useful for detecting those compounds that are non-ionic, do not absorb ultraviolet light and do not fluoresce. E.g. sugar, alcohol, fatty acid and polymers. But it has a Disadvantages: Lack of high sensitivity, Lack of suitability for the gradient elution, Need for strict temperature controlling.

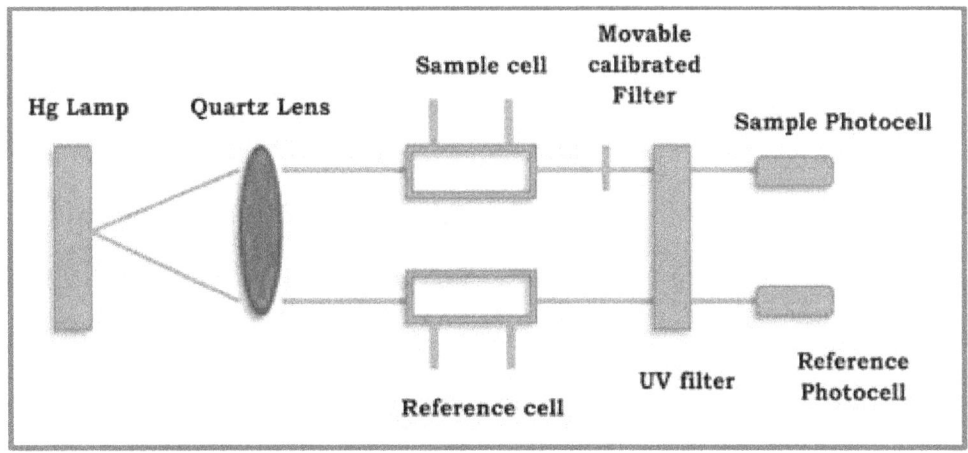

**Figure 2.16 Schematic diagram of a RI detector**

### 2.14.4 Fluorescence Detector

A UV/UV-VIS detector monitors the absorption of light with a specified wavelength. However, some substances absorb light at one wavelength, and then emit light called fluorescence at another wavelength. These detector are both selective and sensitive for those materials which are naturally fluorescence or materials can be made florescent by derivatization. Sensitivity and limit of detection of these detector is extremely high as compared to absorbance detector.

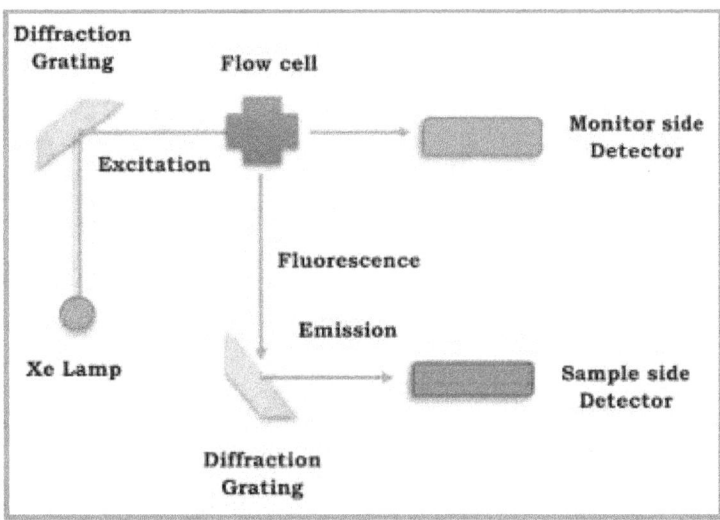

**Figure 2.17 Schematic diagram of a Fluorescence detector**

Fluorescence detection is suitable for trace analysis. These detector is useful for the detection of the natural fluorescence as well as the compounds do not emit fluorescence such as amino acids, etc. can be detected as fluorescent substances, after reaction with a fluorescence reagent (**derivatization**). This method makes it possible to measure various components with high sensitivity. Derivatization can be carried out either before the separation (pre-column) or after the separation (post-column). **Pre-column** derivatization does not requires modification in instruments will **Post-column** derivatization requires a special reactor is situated between column and detector. Non-fluorescent reagent are called **flurotags**. For example Dansyl chloride is used to obtain fluorescent derivatives of proteins, ammines and phenolic compounds.

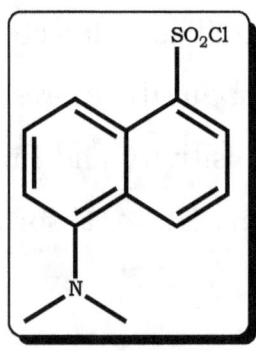

**5-dimethylamono-napthelene-1-sulphonyl chloride**

# Chapter 3 - Liquid chromatography-mass spectrometry (LC-MS)

## 3.1 Introduction

LC/MS is a powerful analytical techniques that combines the resolving power of liquid chromatography with the detection specificity of mass spectrometry. LC separates the sample components and then introduces them to MS. MS creates and detects charged ions. The LC/MS data may be used to provide the information about the molecular weight, structure, identity, quantity, and purity of a sample. Mass spectral data add specificity that increases confidence in the results of both qualitative and quantitative analyses. LC/MS systems facilitate the analysis of samples that traditionally have been difficult to analyse. Despite the power and usefulness of gas chromatography/mass spectrometry (GC/MS), many compounds are impossible to analyse with GC/MS due to the unsuitability for polar, larger, ionic, non-volatile and thermally fragile (thermally unstable) molecules while LC/MS based methods can be applied for wide range of the compounds from small organic molecules to larger proteins.

## 3.2 Importance of MS in LC or sensitivity and selectivity

For most of compounds, a mass spectrometer is more sensitive and more specific than all other LC detector. It can be more selective when detection and identification of components in complex matrix and impurities at trace levels. Using MS in combination with other LC detectors gives richer information. For example, a DAD acquires data on selected ultraviolet (UV) and visible (Vis) wavelengths. This information is useful for identifying unknown peaks and for determining peak purity or for both but does not give the information related to the structure of the molecule. A mass give the information by detecting the ions and offers molecular weight and structural information. It can analyse compounds which does not having chromophore group (UV transparent). It can also identify the components having a same UV spectra or similar mass. It can identify the molecular structure of the unresolved chromatographic peaks and reducing the need of perfect chromatography.

## 3.3 Instrumentation

LC/MS systems have a chromatographic unit that can be used to separate the components and generate the chromatographic peaks. A mass spectrometers ionizes the molecules and then sorting and identify the ions according to their mass to charge ratio. Two key components in this process are the ion source, which generates the ions, and the mass analyser, which sorts the ions. Several different types of ion sources and mass analysers are commonly used for LC/MS. Each is suitable for different classes of compounds. Each has advantages and disadvantages depending on the type of information needed.

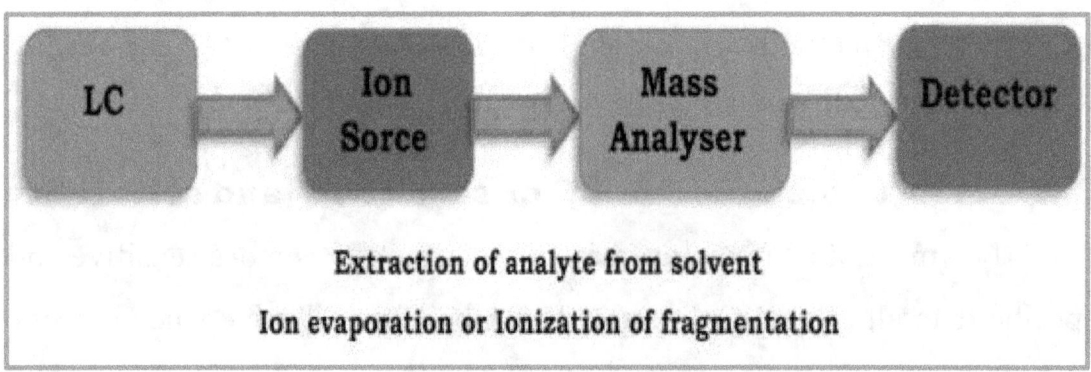

**Figure 3.1 Schematic diagram of LC-MS components**

## 3.4 Interfacing LC and MS

There has been a major focus on improving the interface between the LC and the MS. Liquid chromatography uses high pressure to separate the components in **liquid phase** and produces a high gas load.

Mass spectrometry requires a **vacuum** and a limited gas load. For example, common flow from an LC is 1 ml/min of liquid, when it is converted to the gas phase, is 1 l/min. A typical mass spectrometer can accept only about 1 ml/min of gas.

Furthermore, an LC operates at near ambient temperature where as an MS requires an elevated temperature. There is no mass range limitation for samples analysed by the LC but there are limitations for an MS analyser. Finally, LC can use inorganic buffers and MS prefers volatile buffers.

## 3.5 Ion Sources

Earlier in LC/MS systems, traditional electron ionization used as interfaces but it was applicable only for a very limited number of compounds. The introduction of atmospheric pressure ionization (API) techniques greatly expanded the number of compounds that can be successfully analysed by LC/MS. In atmospheric pressure ionization, the analyte molecules are ionized first, at atmospheric pressure. The analyte ions are then mechanically and electrostatically separated from neutral molecules.

Common atmospheric pressure ionization techniques are:
1) Electrospray ionization (ESI)
2) Atmospheric pressure chemical ionization (APCI)
3) Atmospheric pressure photoionization (APPI)

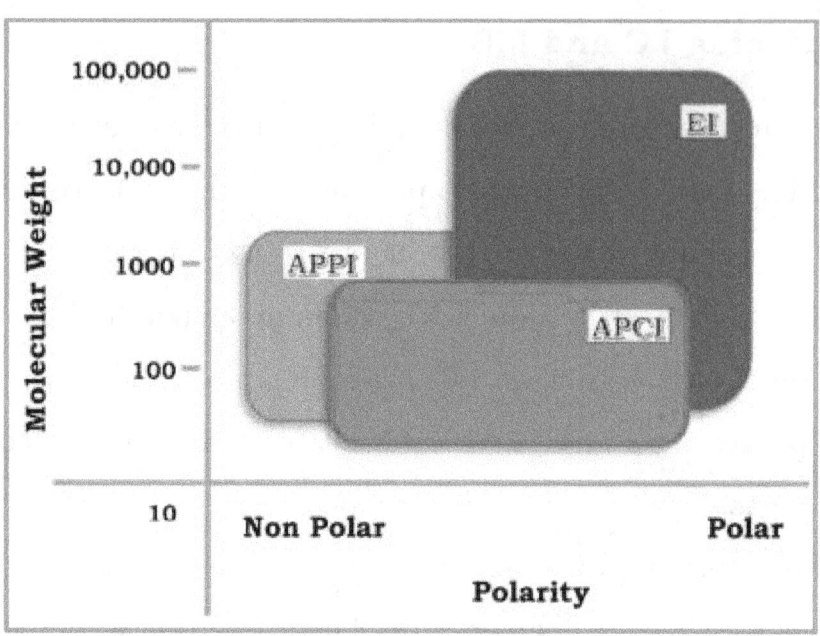

**Figure 3.2 Application of API techniques for compounds on the basis of polarity and molecular weight**

### 3.5.1 API-ES (Electrospray ionization)

API-ES is a process of ionization followed three basic steps: (1) Nebulization and charging (2) Desolvation and (3) Ion evaporation.

The HPLC effluent is sprayed (nebulized) through capillary which is at ground potential. Then spray goes through a semi cylindrical electrode which is at a high potential. The potential difference between the capillary and electrode produces a strong electrical field. Due to the strong electrical, charged droplets produces. Uncharged droplets can be remove by the counter flow of heated nitrogen drying gas which also shrink the droplets. As the droplets shrink, the charge concentration in the droplets increases. They approaches a point where a columbic force (repulsion) exceed the cohesive force. This process continue until the analyte ions are ultimately converted into the gas phase ions which is then pass through the mass analyser.

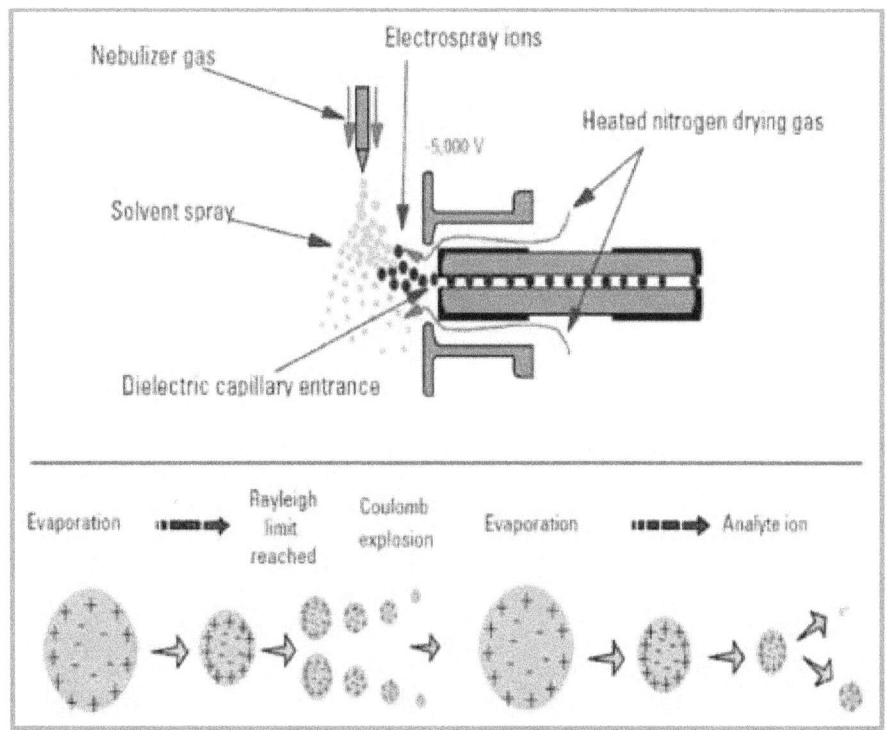

**Figure 3.3 Schematic diagram of ES process in positive mode**

Electrospray is especially useful for analysing large biomolecules such as proteins, peptides, and oligonucleotides because large molecules often acquire more than one charge.

## 3.5.2 APCI (Atmospheric Pressure Chemical ionization)

APCI nebulization is similar to that is API-ES. The HPLC effluent is sprayed (nebulized) through capillary into hot (typically 250-400 °C) vaporization chamber. The heat rapidly evaporates the spray droplets and converting them to gas phase ions of solvent and analyte. These gas phase ions are ionized by the corona discharge needle. The gas phase solvent molecules transfer its charge to the gas phase analyte molecule through a chemical reactions. Then these analyte ions are travel toward mass analyser.

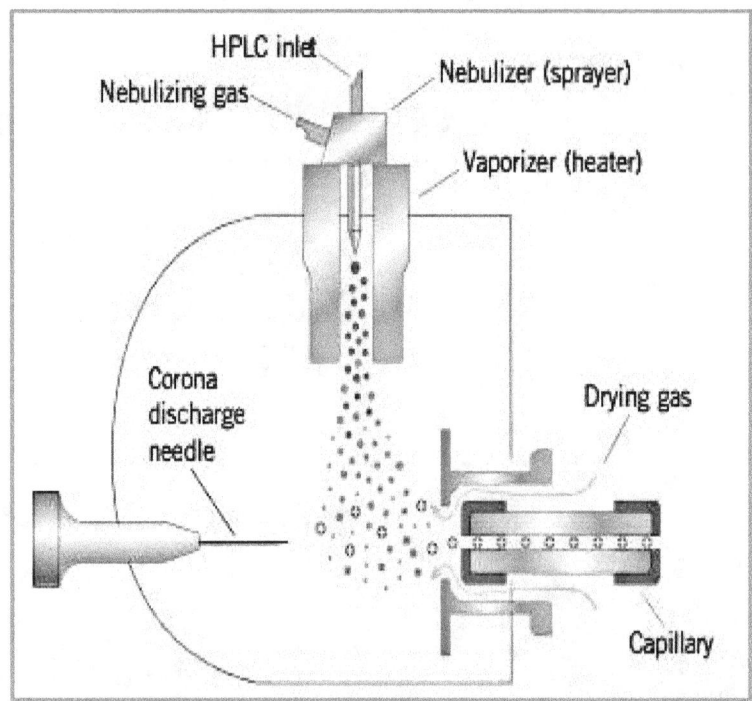

Figure 3.3 Schematic diagram of APCI

APCI is applicable to a wide range of polar and nonpolar analytes that have a moderate molecular weights. APCI is less suitable than the API-ES for the analysis of large biomolecules that may be thermally unstable because it involves high temperature.

## 3.5.3 APPI (Atmospheric Pressure photo ionization)

Similar to APCI, HPLC effluent is sprayed and vaporized into hot vaporization chamber to convert the effluent into gas phase. These gas phase ions are ionized by the UV lamp generate a photos in a narrow range of ionization energies. The range of carefully chosen to ionize analyte molecule maximum as possible while minimizing the ionization of solvent molecules. The resulting ions are then travel towards mass analyser.

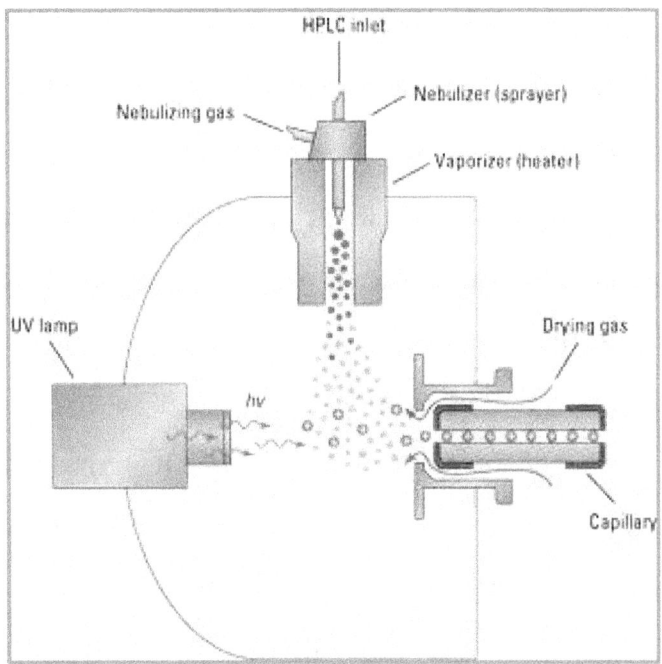

**Figure 3.4 Schematic diagram of APCI**

## 3.6 Mass Analysers

Although in theory any type of mass analyser could be used for LC/MS, four types:

1. Quadrupole
2. Time-of-flight
3. Ion trap are used most often.

Each has advantages and disadvantages depending on the requirements of a particular analysis.

### 3.6.1 Quadrupole Mass analyser

The quadrupole mass spectrometer is a scanning mass analyser that uses the stability of the ion trajectories in oscillating electrical field to separate them according to their mass to charge ratios (m/e). It consists of four parallel metal rods to which both DC voltage and an oscillating RF voltage is applied. Two opposite poles are positively charged and other two negative charged. And their polarities change by alternating the voltages throughout the experiment. As ions from the ionization source enter the RF field along the z axis of the electrodes. Ions having a high m/e ratio, will less deflected along the RF filed and have a stable path through the detector. Others having low m/e ratio, will more deflected along the RF field (unstable path) and collide with the electrodes and be lost. This is called high pass mass filtration for the lager m/e ions. For the selection of the smaller m/e ions, DC current is applied along with the RF voltage, lager particles are deviated more and take more time to hit the detector while smaller particles are deviated less and take a less time to hit the detector. This is called low pass mass filtration. Quadrupole mass analyzers can operate in two modes:

- Scanning (scan) mode
- Selected ion monitoring (SIM) mode

In scan mode, the mass analyzer monitors a range of mass-to-charge ratios. In SIM mode, the mass analyzer monitors only a few mass to-charge ratios.

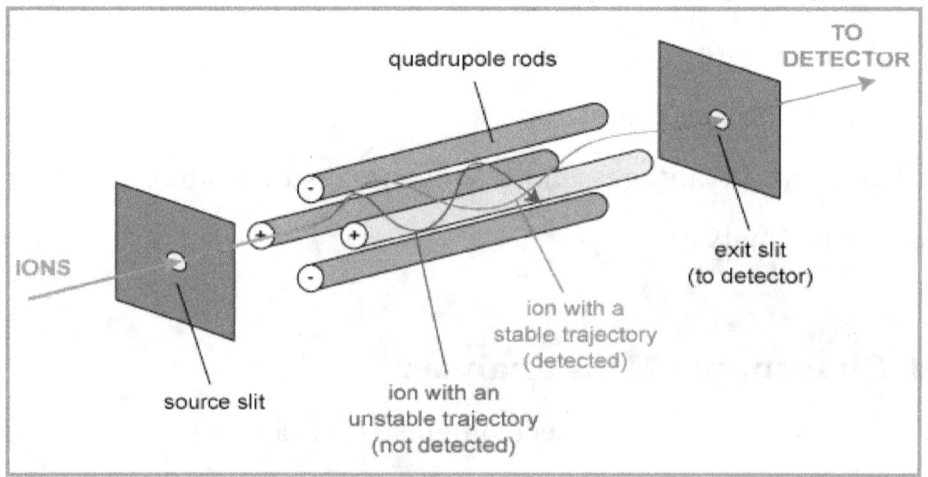

**Figure 3.5 Schematic diagram of quadrupole mass analyser**

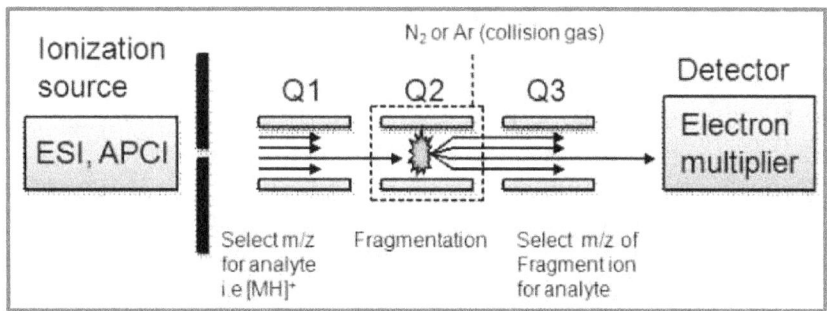

**Figure 3.6 Schematic diagram of triple quadrupole mass analyser**

## 3.6.2 Time of flight (TOF)

A time of flight mass spectrometer is a non-scanning mass analyzer similar to chromatography, expect there are no stationary phase /mobile phase, instead the separation is based on the kinetic energy and velocity of the ions. In TOF, ions formed in the ionization chamber are accelerated through the repelled plates and then enter into a flight tube with constant kinetic energy (ions of the same charge have equal kinetic energies). The ions of different m/e travel at different velocities. The kinetic energy of ions leaving the source is given by

$$KE = zV = \frac{mv^2}{2} \quad \ldots\ldots\ldots\ldots\ldots\ldots (3.1)$$

Where m = mass of ion, v = velocity of ion, V = accelerating voltage, z = ion charge

Rearranging

$$v = \sqrt{\frac{2Vz}{m}} \quad \ldots\ldots\ldots\ldots\ldots\ldots (3.2)$$

The time of flight or time it takes for the ion to travel the length of the flight tube is

$$T = \frac{L}{v} \quad \text{............................(3.3)}$$

Where L = length of tube and v = velocity of ion

Substituting equation (3.2) for kinetic energy in equation (3.3) for time of flight

$$T = L\sqrt{\frac{m}{z}\frac{1}{2V}} \;\alpha\; \sqrt{\frac{m}{z}} \quad \text{............................(3.4)}$$

During the analysis, L and V are held constant so that time of flight is directly proportional to the root of the mass to charge ration. Lighter ions travel faster and arrive at the detector first, so the mass-to-charge ratios of the ions are determined by their arrival times. TOF have a wide mass range and can be very accurate in their mass measurements.

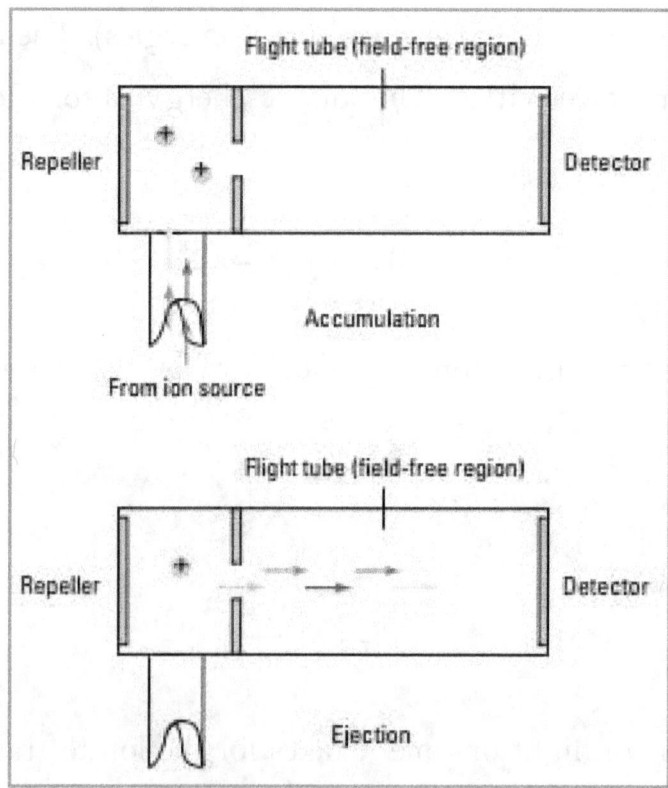

**Figure 3.7 Schematic diagram of TOF mass analyser**

### 3.6.3 Ion trap

An ion trap mass analyzer consists of a circular ring electrode plus two end caps that together form a chamber. Ions entering the chamber are "trapped" by electromagnetic fields. Another field can be applied to selectively eject ions from the trap. Ion traps have the advantage of being able to perform multiple stages of mass spectrometry without additional mass analyzers.

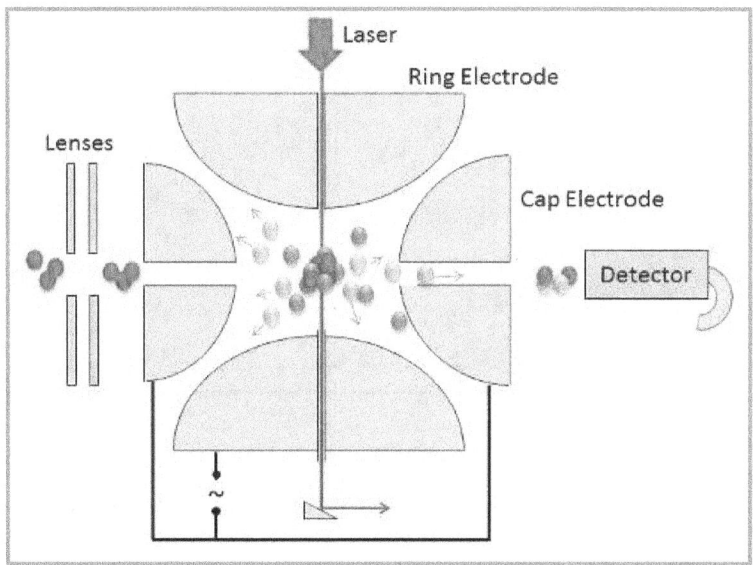

**Figure 3.8 Schematic diagram of Ion trap mass analyser**

## 3.7 Applications

LC-MS is a techniques that is widely used in basic research, pharmaceutical, food and other industries. Some of the applications are mention below.

## 3.7.1 Molecular weight determination

The spectra of Octa peptides whose m/z ratio differ only by 1 m/z. The only difference in sequence is at C-terminus that is one having threonine and other having threonine amide. The smaller fragments are identical in the two spectra, indicating that large portions of the two peptides are very similar. The larger fragments contain the differentiating peptides.

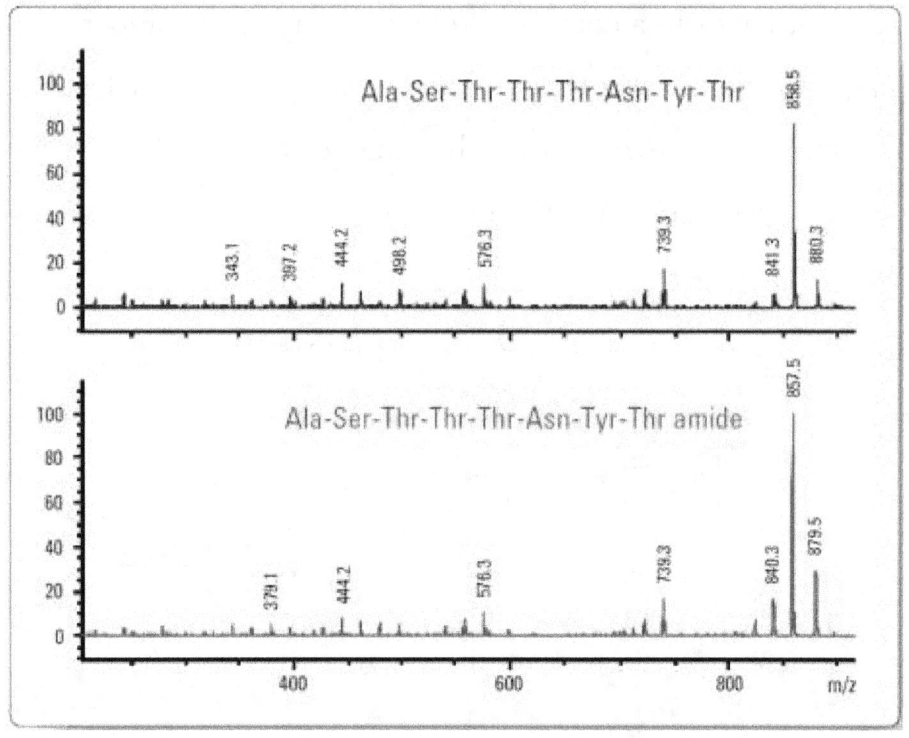

Figure 3.9 Mass fragmentation of threonine and threonine amide

## 3.7.2 Pharmaceutical Applications (Rapid chromatograph of benzodiazepines)

It allows compounds to be separated even they are chromatographically unresolved. A series of benzodiazepines are analyzed using both UV and MS detectors. In this chlorine $Cl^-$ has a characteristic abundance of 2 most abundant isotopes. By which TRAIZOLAM spectrum shows a two $Cl^-$ ions and DIAZEPAM shows only one $Cl^-$ ion.

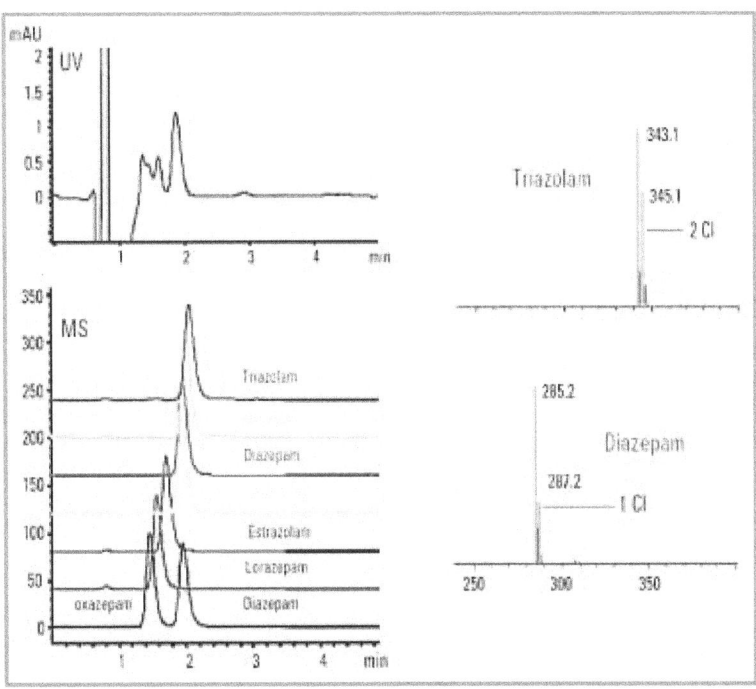

**Figure 3.10 Rapid Chromatogram of a series of benzodiazepines and isotopic separation of triazolam and diazepam**

## 3.7.3 Food applications (Identification of Aflatoxins in food)

Aflatoxins are fungi produced in food. By total in chromatogram we can detect 4 different aflatoxins. Even though they are structurally very similar to each other but can be uniquely identified by LC/MS.

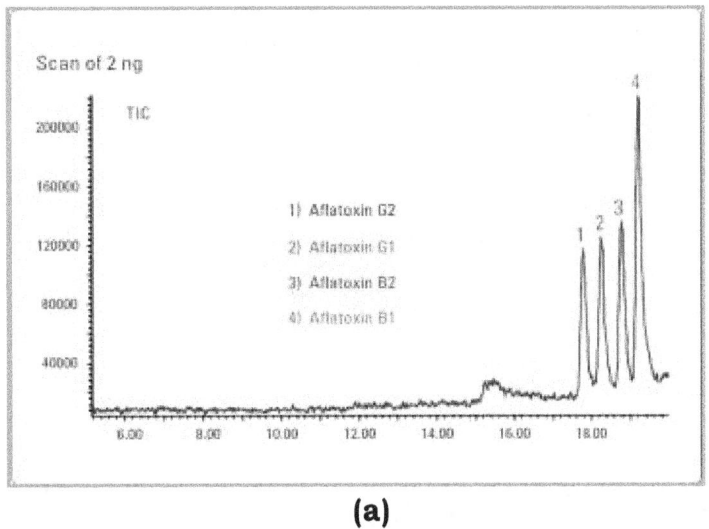

(a)

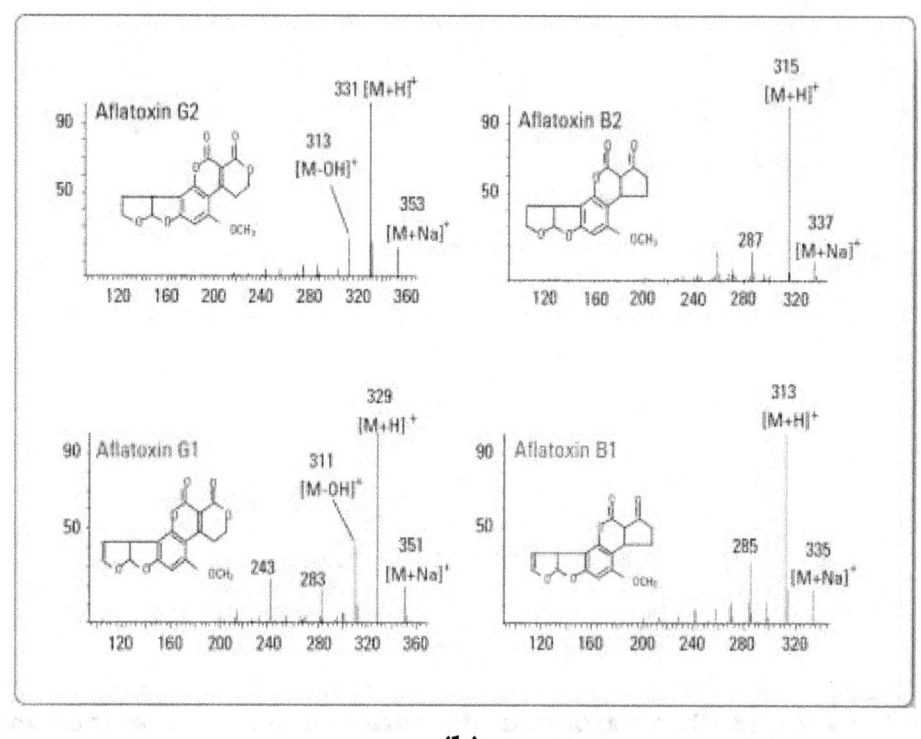

(b)

**Figure 3.11 a) Chromatogram of four different aflatoxin fungi species and b) Mass spectrum of aflatoxin fungi species present in food**

# Chapter 4 - Gas Chromatography (GC)

## 4.1 Introduction

"GC is a method of fractioning the components of the vaporized sample as a consequence of being partitioned between a mobile phase and a stationary phase held in a column"

There are two type of gas chromatography: Gas-Liquid chromatography (GLC) and Gas-Solid chromatography (GSC). GLC has a wide application in the field of science. It is generally known as GC. The primary limitation of GC is that the sample must volatized without undergoing decomposition (thermally stable). This limitation is replaced by HPLC. But the faster separation and higher efficiency is observed as compared to HPLC due the high flow rate of mobile phase and plate heights (longer column length).

## 4.2 Difference of GSC and GLC

In GSC, Stationary phase is solid and mobile phase is gas and separation is based on physical adsorption of gaseous substance on solid surface. In GLC, Stationary phase is liquid coated on the inert solid packing and mobile phase is gas and separation is based on partition between phases. In GSC, distribution constants are generally much larger than GLC. As a result, GSC is useful for the species that are not retained by GLC such as components of air, carbon monoxide, carbon dioxide, nitrogen oxides and rare gases. GSC has a limited application because of the semi-permanent retention of polar molecules leads tailing of elution peaks. So this techniques is not useful except for the separation of certain low molecular weight gaseous species.

## 4.3 Instrumentation

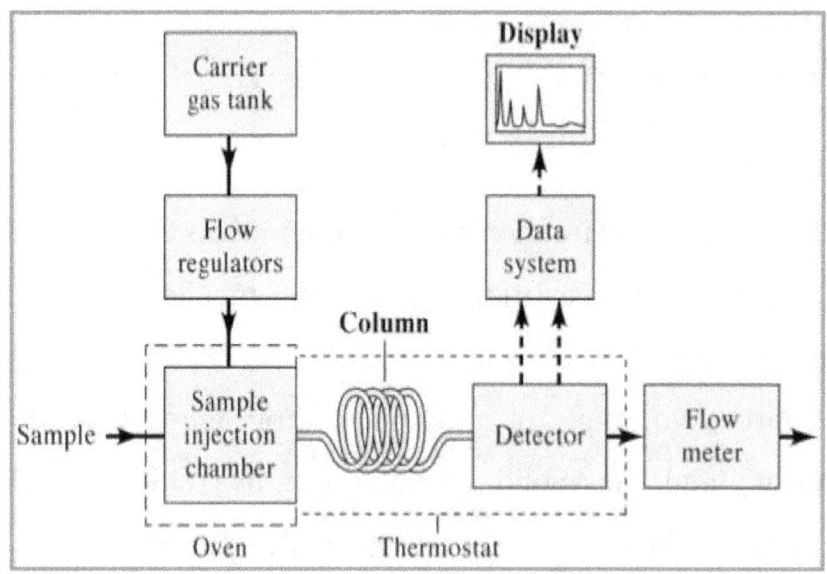

Figure 4.1 Schematic Diagram of Gas Chromatography

### 4.3.1 Carrier Gas system

The mobile phase gas in GC is called a carrier gas and it must be inert chemically, suitable for the detector, give best column performance, not cause risk of fire & explosion. Generally, Helium is widely used as carrier gases. Some of other gases are hydrogen, helium, nitrogen, air and argon. These gases are available in pressurized tanks. In addition, the carrier gas system often contains a molecular sieve to remove water or other impurities.

### 4.3.2 Flow regulators and flow meter

Pressure regulators and flow meters are required to control the flow rate of the gas. Flow rate is in between 25 to 150 ml/min with packed columns and 1 to 25 ml/min for open tubular capillary columns. Usually **Soap-bubble flow meters** is placed at the end of column to measure flow rate.

A soap film is formed in the path of the gas when a rubber bulb containing an aqueous solution of soap and detergent is squeezed. The time required for the movement of the film between two graduations on the burette is measured and converted to volumetric flow rate. The linear relationship between volumetric flow rate and linear flow velocity

$$V_R = t_R F \ldots\ldots\ldots\ldots\ldots(4.1)$$

Here, F = flow rate of gases mobile phase, $V_R$ = volume of mobile phase between upper marks and lower marks, $t_R$ = time required to reach mobile phase from lower to upper point.

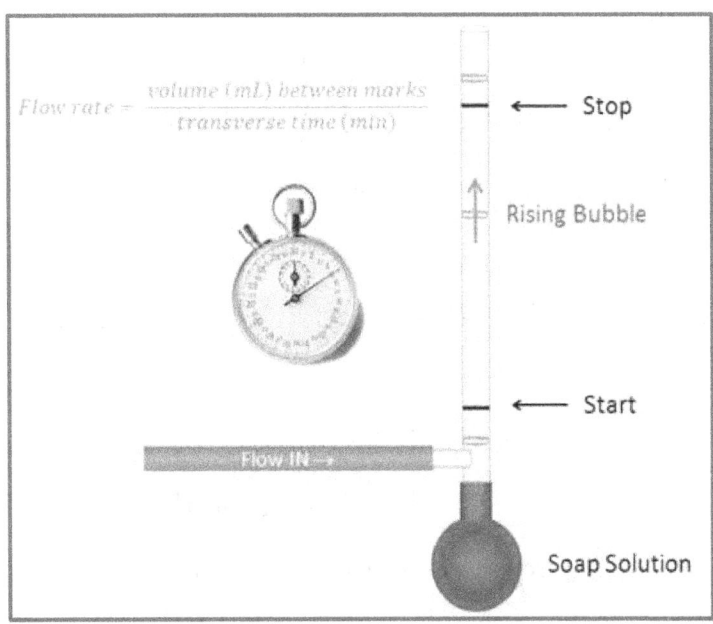

Figure 4.2 Schematic Diagram of Bubble flow meter

### 4.3.3 Sample injection system

To achieve high column efficiency requires that the sample must be in suitable size. Slow injection or oversized samples causes band spreading and poor resolution. Generally micro syringes used to inject liquid or gaseous samples through a self-sealed rubber septum.

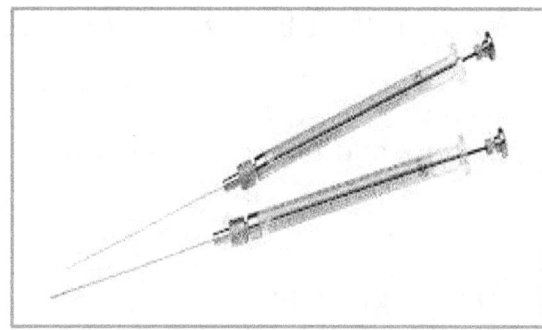

**Figure 4.3 Schematic Diagram of Micro syringe with 1μL to 20 μL used for sample injection in GC**

Solid sample are dissolved in volatile solvents or injected directly if they can be liquefied. Gas samples requires special gas sampling valves. For ordinary analytical column, 20 μL sample size is used. Two type of sample injection technique is used for the injection of sample.

**Sample splitter** – Small known fraction of sample is delivered and remaining going to waste.

**Spilt less injection** - With split less inlets, the purge valve close at injection and stay closed for 0-60 sec. During this time the sample vapor can go only on the column. When purge valve opens, remaining vapor is rapidly vented. Rotation of the valve by 45° then introduces the reproducible volume ACB into the mobile phase.

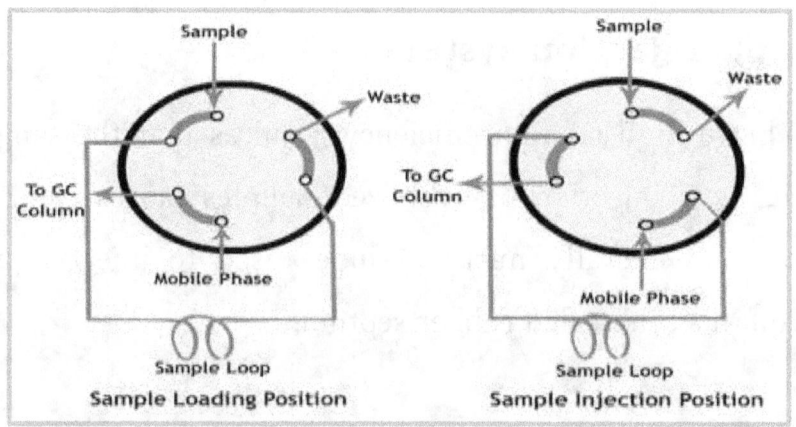

**Figure 4.4 Schematic Diagram of position of valve**

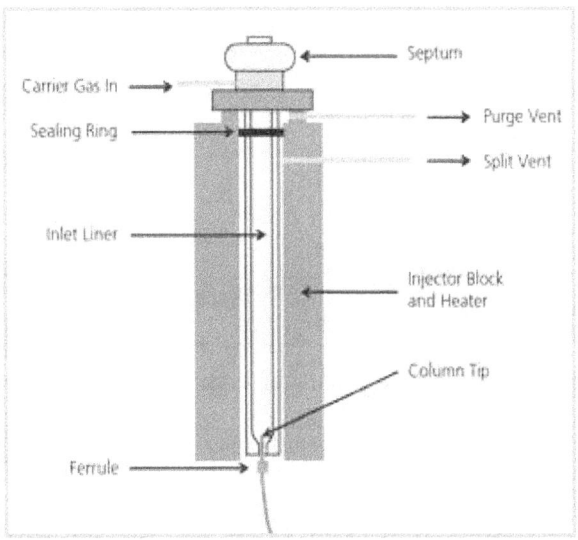

Figure 4.5 Schematic Diagram of spilt/spiltless injection

## 4.3.4 Column Configurations

Two general types of columns are encountered in gas chromatography, **packed** and **open tubular**, or **capillary**.

**Packed column-** length from 1m to 5 m.

**Capillary column-** length from a few meters to 100 m.

In the past, the majority of GC analysis packed column was used. For most current applications, packed column have been replaced by the capillary column. They are constructed of stainless steel, glass, fused silica, or Teflon. They are usually formed as coils having diameters of 10 to 30 cm.

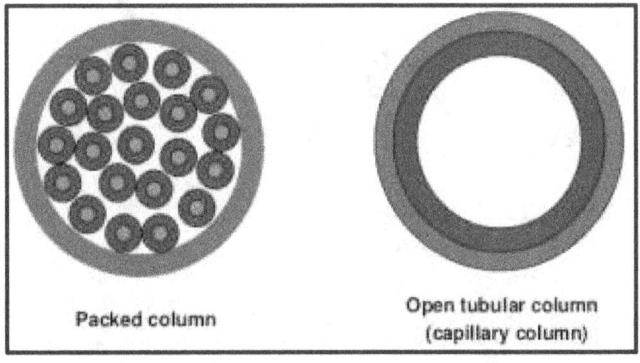

Figure 4.6 Schematic Diagram of packed and capillary column

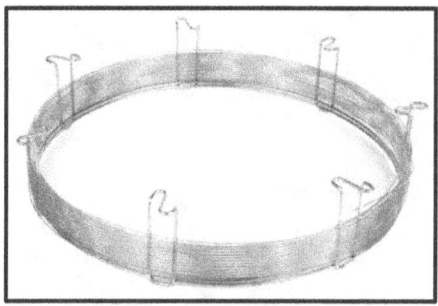

**Figure 4.7 Schematic Diagram of capillary column with 100 m long**

**Open tubular column or capillary column** are three basic type

1. **Wall-coated open tubular (WCOT)** – simple capillary tube made by glass and inside wall of the tube is coated with a liquid phase.

2. **Support-coated open tubular (SCOT)** – Columns packing having a micron size support coated with a thin film of the liquid phase. Support material is such as diatomaceous earth (calcite celite-firebrick and glass beads). These column have more sample capacity and inlet splitter not be required. These columns are preferred for the trace analysis.

3. **Porous layer open tubular column (PLOT)** - simple capillary tube made by glass and inside wall of the tube is coated with a solid phase particles. The difference between SCOT and PLOT is that the PLOT does not have a liquid stationary phase.

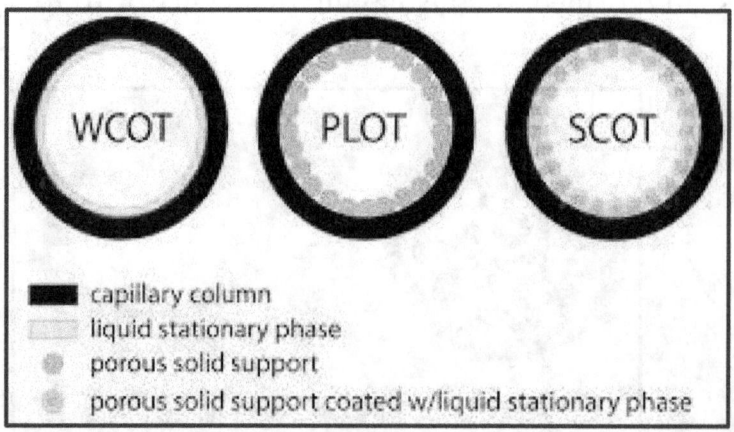

**Figure 4.8 Schematic Diagram of cross section of capillary column**

### 4.3.5 Column temperature (temperature programming)

Column temperature is an important variable. It must be controlled to a few tenths of a degree for precise work. Thus, the column is ordinarily housed in a thermostat oven. The optimum column temperature depends upon the **boiling point of the sample** and the **degree of separation required**. Roughly, a temperature equal to or slightly above the average boiling point of a sample results in a reasonable elution time (2 to 30 min). For samples with a broad boiling range, it is often desirable to employ temperature programming, whereby the column temperature is increased either continuously or in steps as the separation proceeds.

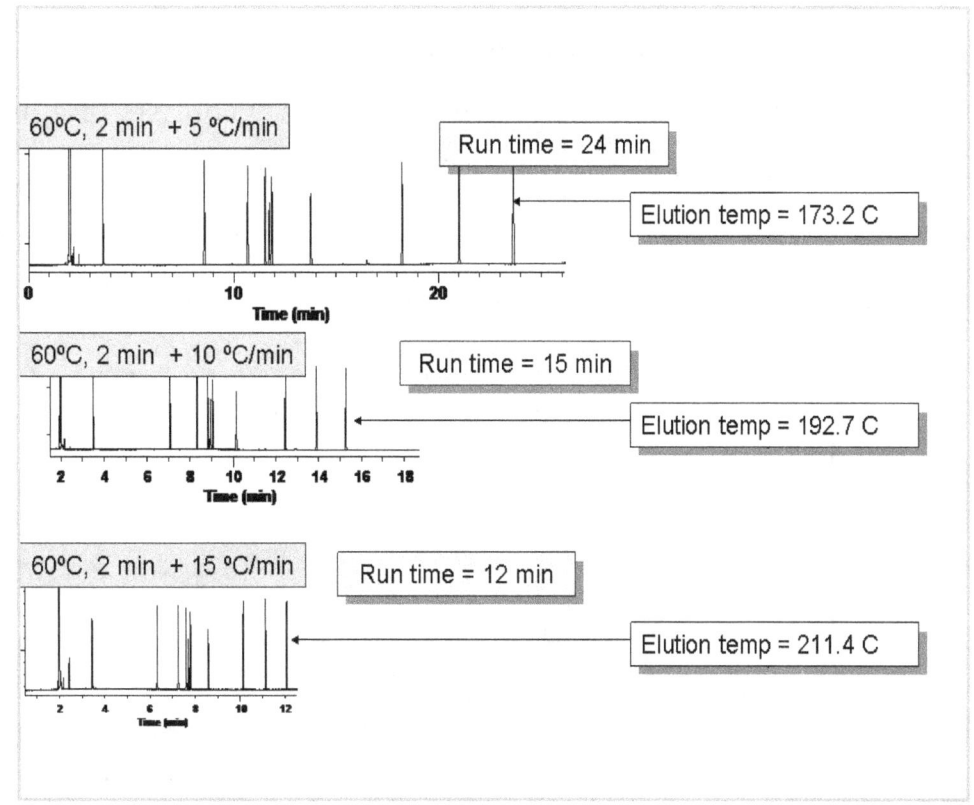

**Figure 4.9 Temperature programming –Changing the temperature of column with time to simulate gradient elution in GC**

## 4.4 Detection System

All the detector monitor the GLC column effluent by measuring the changes in the composition arising from variations in the eluted components. When carrier gas alone is passing they give a zero signal. When a component is eluted, it is detected and signal proportional to the amount of components.

Most widely used detectors used in GC are Thermal conductivity detector (TCD), Flame ionization detector (FID) and Electron capture detector (ECD).

## 4.4.1 Flame Ionization Detectors (FID)

Application is based upon changes in the electrical conductivity of the gas stream brought by the presence of analyze molecules. At normal temperature and pressure, gases act as insulator it does not ionized but it will become conductive of ions if electrons are present. An FID consists of a hydrogen/air flame and a collector plate (electrode). The effluent from the column passes through the flame which breaks down organic molecules and produces ions. The ions are collected on based collector electrode and produced an electrical signal.

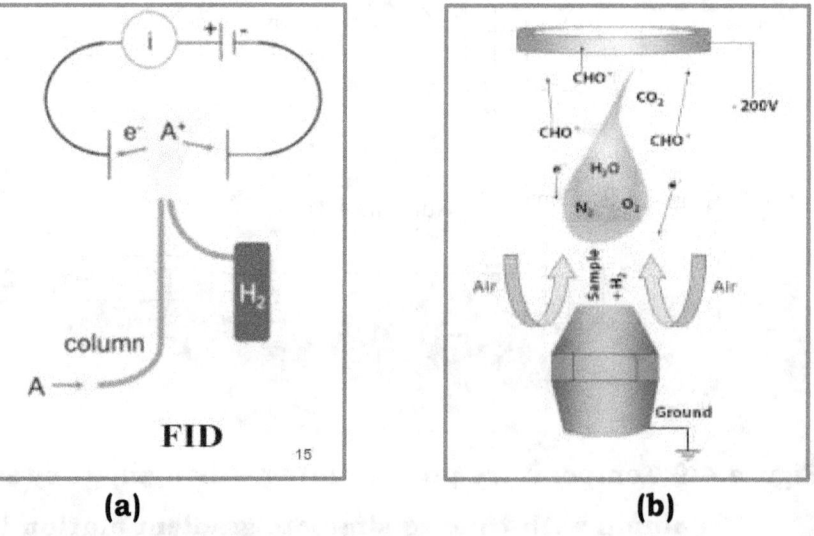

(a)            (b)

**Figure 4.10 (a) Schematic diagram of FID in GC (b) Sample are combusted in flame, creating positive ions and electrons. The positive ions are attracted to negatively biased collector while electrons repelled towards jet**

The ionization of carbon compounds in the FID is not fully understood, although the number of ions produced is roughly proportional to the number of reduced carbon atoms in the flame. So **it is called mass sensitive device** rather than a concentration sensitive. The detector is insensitive towards noncombustible gases such as $CO_2$, CO, $SO_2$, noble gases and $NO_x$. A disadvantage of the flame ionization detector is that destroy the sample during the combustion step.

## 4.4.2 Thermal Conductivity Detector (TCD)

It was one of the earliest detector for GC, is still widely used. Application is based upon changes in the thermal conductivity of the gas stream brought by the presence of analyze molecules. Detector consist of an electrically heated filaments (platinum, gold or tungsten) with an applied current. The temperature of element depends upon the thermal conductivity of the surrounding gas. When the carrier gas passes in the both the cell, temperature of the sample cell changes due to wide difference in thermal conductivity of sample than carrier gas. As a result the resistance through which current is flowing also varies and signal is generated.

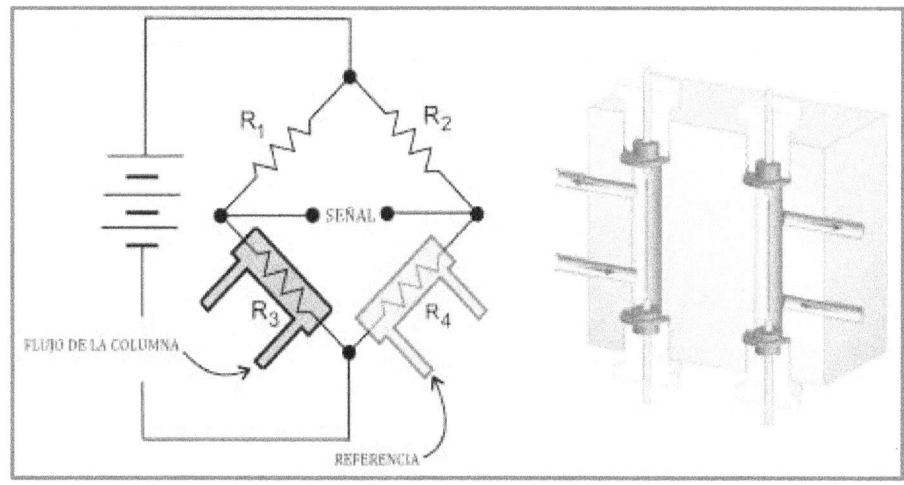

**Figure 4.11 Schematic diagram of TCD detection**

### 4.4.3 Electron-Capture Detector (ECD)

The ECD detector is highly sensitive detector that widely used in the analysis of detection organchlorine pesticides, hazardous halide substances etc. These detector is worked based on the capture of electrons by electronegative atoms in molecules. Electrons are produced by ionization of the carrier gas with radioactive nickel-63 which produced a current between a based pair of electrodes. In the absence of organic species, a constant standing current is produced, while compounds with electronegative atoms capture electrons and reducing a current.

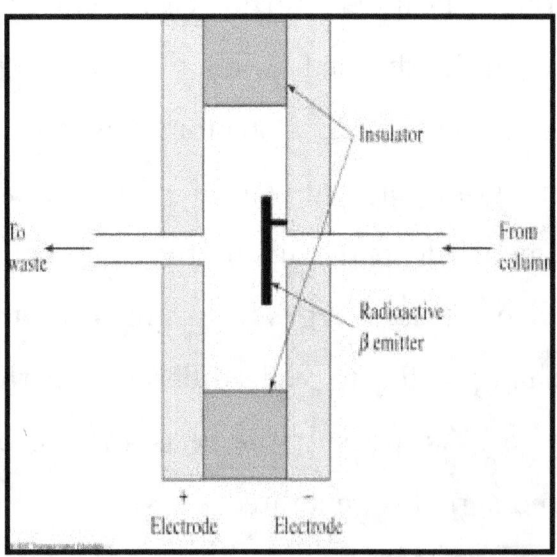

**Figure 4.11 Schematic diagram of ECD**

# Chapter 5 - Gas Chromatography - Mass Spectrometry (GC-MS)

## 5.1 Introduction

Gas Chromatography Mass Spectrometry (GC-MS) combines two powerful techniques to provide the identification of compounds with low detection limits and the potential for quantitative analysis. GC-MS analyses can work on liquid, gaseous and solid samples, but is primarily restricted to volatile and semi-volatile compounds.

## 5.2 Instrument

In GC-MS, a sample is volatilized and carried by an inert gas through a coated glass capillary column. The time it takes a specific compound to pass through the column to a detector is called its "retention time", which can be used for identification when compared to a reference.

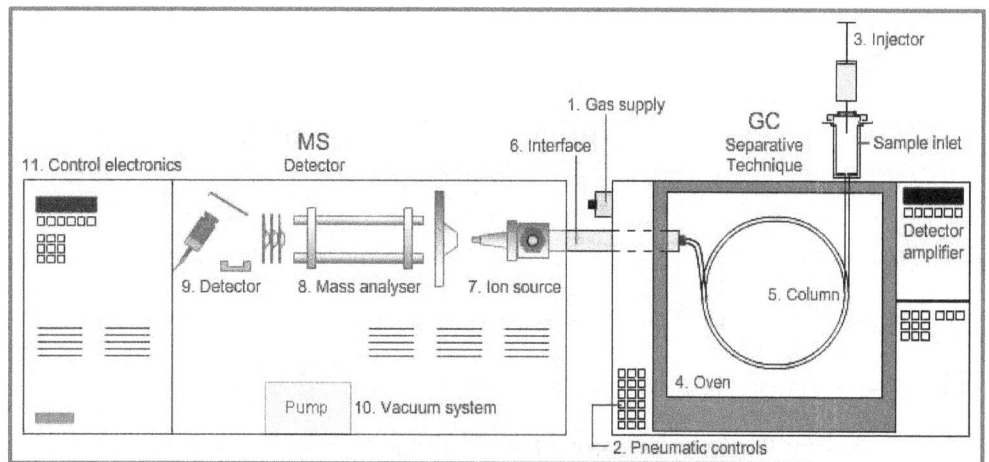

**Figure 5.1 Schematic diagram of GC-MS**

Compounds leaving the GC column are fragmented by electron impact. The charged fragments are detected, and the subsequent spectrum obtained can be used to identify the molecule. Fragmentation patterns are reproducible, and can be used to produce quantitative measurements.

## 5.3 Sample Used

**GC-MS can be performed on liquids, gases or solids.** For liquids and gases the sample is commonly directly injected into the GC. For solids, the analysis is carried out by solvent extraction, outgassing (desorption) or pyrolysis.

### 5.3.1 Sampling Techniques

In GC-MS, liquids and gases the sample is directly injected into the GC while for solids, the analysis is carried out by solvent extraction, outgassing (desorption) or pyrolysis.

Desorption experiments are performed under helium flow at a controlled temperature between 40-300 °C, with analytes being collected on a cryogenic trap during desorption.

Pyrolysis is another sampling technique for the analysis of materials that cannot be directly injected into the GC-MS. By applying heat directly to the sample, the molecule can be broken down in a reproducible way. The smaller molecules are then introduced into the GC and analysed by GC-MS. By this method, probe temperatures of up to 1400°C can be used.

Many other sample preparation and sampling methods are available, such as derivatization, static headspace analysis, purge & trap, SPME (solid phase micro extraction) that have applications based on the sample type and species of interest.

## 5.4 Interface or coupling of GC-MS

After separation in the GC, capillary column outlet needs to be connected to the ion source of the mass spectrometer which need to fulfil the following conditions:

- Analyte must not condense in the interface
- Analyte must not decompose before entering the mass spectrometer ion source
- The flow rate of mobile phase must be within the pumping capacity of the mass spectrometer.

Most common capillary GC columns have an internal diameter below around 0.32 mm will not require interface due to the flow rate associated with these columns is low. However columns having an internal diameter around 0.53 mm used will require special interfacing with MS due to its produces a high flow rate. The flow rate of mobile phase in GC-MS do not exceed 2.0 mL/min can usually be achieved by using direct interfaces. At flow rates above around 2mL/min will require the use of vapour concentrator devices (like the Bieman concentrator) or jet separator interfaces. Even the most efficient two stage vacuum systems will struggle to attain the required level of vacuum ($\sim 10^{-5}$ to $10^{-6}$ torr) to carry out the analysis. Further, filaments will have much reduced lifetimes at compromised vacuum levels.

Interface that commonly used in GC and Ms Coupling are **Jet separator, Bieman concentrator and Direct introduction,**

### 5.4.1 Jet separator Interface

GC columns produce a relatively high volumetric flow of carrier gas (> 2 mL/min) which if directly introduced into the ion source of mass spectrometer may leading to much reduced efficiency of the ionisation process and sorting the lifetime of filaments. So that the jet separator is used as an interface. This device contains a very small gap between the column and transfer line which is held under vacuum and strip away the lighter carrier analyte molecules form the molecules having a higher mass and travel towards the ion source. The disadvantage is that the some highly volatile analyte molecules may be lost to the vacuum.

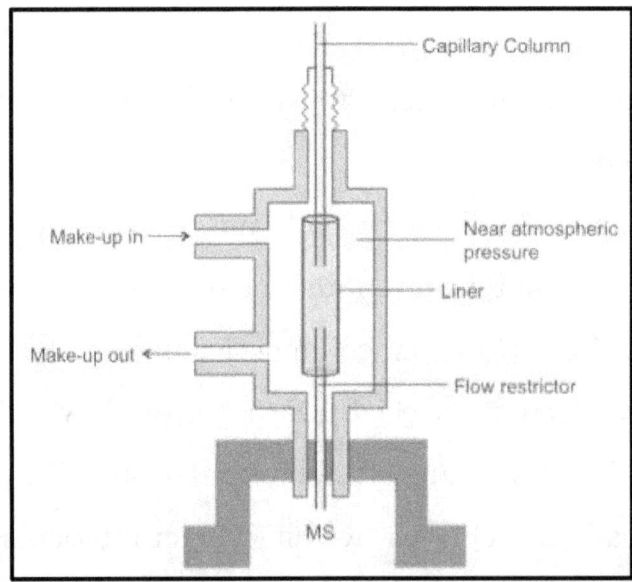

**Figure 5.2 Schematic diagram of Jet Separator Interface**

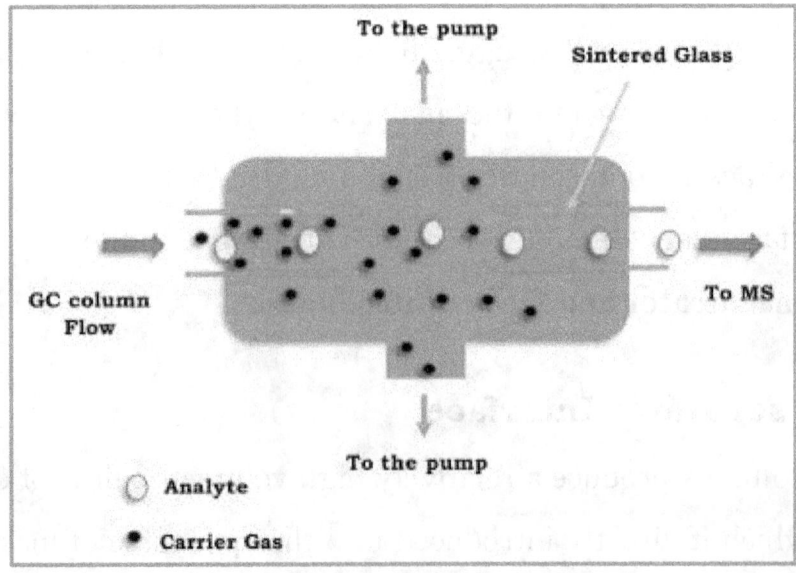

**Figure 5.3 Schematic diagram of Jet Separator Interface**

## 5.4.2 The Bieman Concentrator

The Bieman concentrator device is used with a packed columns or with wide bore capillary columns at high flow rates. The concentrator consists of a heated glass jacket surrounding a sintered glass tube. The eluent from column passes directly through the sintered glass tube and helium diffuses radially through the porous wall and is continuously pumped away. The helium stream enriched with solute vapour passes on to the mass spectrometer.

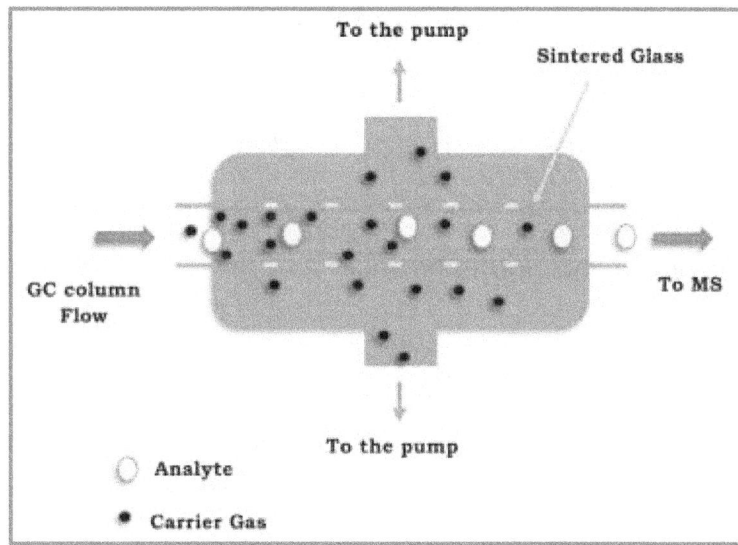

**Figure 5.4 Schematic diagram of Bieman Concentrator**

## 5.4.3 Direct introduction

Direct introduction is typically used with capillary GC column and most modern instruments can easily cope with flow rates up to 2 ml/min. In a direct interface. The column is inserted directly into ionisation chamber.

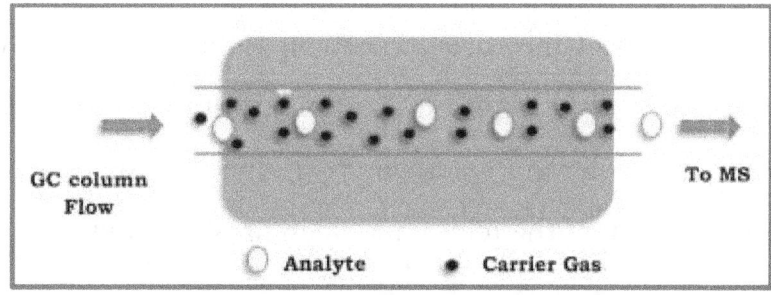

**Figure 5.5 Schematic diagram of Direct Introduction**

## 5.5 Ion Sources

Generally, two types of ion source mostly commonly used in GC/MS are Electron impact ionization (EI) and Chemical ionization (CI).

### 5.5.1 Electron impact ionization (EI)

In EI technique, positive and negative ionization occurs. a high energetic beam of electrons passages to the sample molecules resulting in the ionisation of molecules due to loss of one electron from the highest occupied molecular orbital (HOMO) of an atom and forming a radical cation according to general scheme. A molecule with one electron missing is called the molecular ion ($M^+$). When the resulting peak from this ion is seen in a mass spectrum, it gives the molecular weight of the compound.

$$M \text{ (molecule)} \xrightarrow{70 \text{ eV}} M^{+\cdot} \text{ (molecular ion, a radical cation)} + e^- \text{ (electron)}$$

In general, the order of ease with which electrons are lost under electron ionization conditions is lone pair > π-bonded pair > σ-bonded pair. Due to the large amount of energy imparted to the molecular ion it usually fragments producing further smaller ions with characteristic relative abundances that provide a 'fingerprint' for that molecular structure. This information may be then used to identify compounds of interest and help elucidate the structure of unknown components of mixtures. Electron ionization is known as a "harsh" technique because of the degree of fragmentation induced. It is possible to reduce the energy of ionizing electrons to reduce the degree of fragmentation, however the sensitivity of the technique decreases markedly.

## 5.5.2 Chemical ionization (CI)

Chemical ionization (CI) is a lower energy process than electron ionization because it involves ion/molecule reactions rather than electron removal. The lower energy yields less fragmentation, and usually a simpler spectrum. A typical CI spectrum has an easily identifiable molecular ion.

In a chemical ionization, inside the ion source, the reagent gas is present in large excess compared to the analyte. Electrons entering the source will preferentially ionize the reagent gas. The collisions of ions of the reagent gas with other reagent gas molecules will produced an ionization plasma. Positive and negative ions of the analyte are formed by reactions with this plasma. Some common reagent gases include: methane, ammonia, and isobutane. The mechanism of the formation of predominant reactant ions when methane used as a reagent gas is shown given.

Primary ion formation :

$$CH_4 + e^- \longrightarrow CH_4^+ + 2e^- \longrightarrow CH_3^+ + H^\bullet$$

Reagent ion formation :

$$CH_4^+ + CH_4^+ \longrightarrow CH_5^+ + CH_4$$

$$CH_4^+ + CH3^\bullet \longrightarrow C_2H_5^+ + H_2 + HH$$

Proton Transfer :

$$CH_5^+ + M \longrightarrow CH_4 + [M+H]^+$$

Addition of Reactant ion :

$$C_2H_5^+ + M \longrightarrow [M+C_2H_5]^+$$

In methane positive ion mode chemical ionization the relevant sample peaks observed are $[M+H]^+$, $[M+CH_5]^+$, and $[M+C_2H_5]^+$; but mainly $[M+H]^+$. This corresponds to the masses M+1, M+29, and M+41.

Similarly, reaction mechanism is shown while ammonia is used as reagent gas.

Primary ion formation:

$$NH_3 + e^- \longrightarrow NH_3^+ + 2e^-$$

Reagent ion formation:

$$NH_3^+ + NH_3 \longrightarrow NH_4^+ + NH_2$$

$$NH_4^+ + NH_4^+ \longrightarrow N_2H_7^+ + H$$

Proton Transfer:

$$NH_4^+ + M \longrightarrow NH_3 + [M+H]^+$$

Addition of Reactant ion:

$$N_2H_7^+ + M \longrightarrow [M+N_2H_7]^+$$

In ammonia positive ion mode chemical ionization the main peaks observed are $[M+H]^+$, and $[M+NH_4]^+$. If more than one protonation site is present, additional $NH_3$ adducts might be seen corresponding to $[M+N_2H_7]^+$. This corresponds to the masses M+1, M+18, and M+35. Two factors determine the choice of the reagent gas to be used: Proton affinity PA and Energy transfer

$NH_3$ is the most used reagent gas in CI because of the low energy transfer of $NH4^+$ compare to $CH_5^+$ for example. With $NH_3$ as reagent gas, usually $[M+H]^+$ and $[M+NH_4]^+$ (17 mass units difference) are observed.

## 5.6 Analysers

Mass spectrometers based on magnetic sectors are widely used to determine molecular structure. They deflect ions down a curved tube in a magnetic field based on their kinetic energy determined by mass, charge and velocity. The magnetic field is scanned to measured different ions. This mass separator is very powerful and capable of very high resolution, but the instruments are quit large and not very suitable for the use with GC. Most GC-MS instruments are today benchtop systems that are use more compact, inexpensive mass analysers with lower resolution. These availability of these is the reason GC-MS is so widely used. Although in theory any type of mass analyser could be used for GC/MS, four types:

1. Quadrupole
2. Time-of-flight
3. Ion trap are used most often.

Each has advantages and disadvantages depending on the requirements of a particular analysis. Working principles of each detectors used in GC-MS are define as earlier in LC-MS section.

## 5.7 Difference between LC-MS and GC-MS

Both the techniques are used for separation and identification of the components present in the sample.

### LC-MS

In LC-MS, Mobile phase is a liquid (solvents). Mobile phase is below its critical temperature and above its critical pressure, it acts as a liquid, so the technique is liquid chromatography. LC-MS is mainly used for nonvolatile and thermally unstable molecules in complex sample (Example, analysis of biological fluids) having a molecular weight in Kilo Dalton. LC-MS works on soft ionization techniques. It is useful for the studies purity and impurity profiling in drugs. So these tool has wide application in R & D in pharmaceutical industries.

## GC-MS

In GC-MS, Inert gases (like helium) used as mobile phase. Mobile phase is above its critical temperature and below its critical pressure, it acts as a gas so the technique is gas chromatography (GC). Mainly used for volatile and thermally stable molecules in complex sample (Example, analysis of petroleum products) having a low molecular weights less than 1200 amu. GC-MS works on hard ionization techniques. It useful in petro chemical, pesticides industries and in field of perfumery.

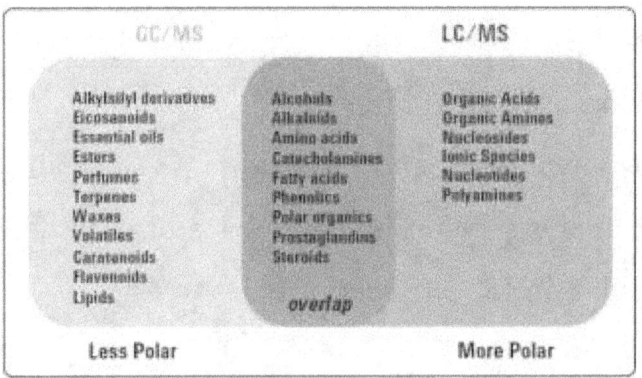

Figure 5.5 Separation of Polar and non-polar via GC-MS and LC-MS

## 5.8 Strength and Limitation of GC-MS

GC-MS is widely used for quantitative analysis as well as identification of organic components by separating complex mixtures. Target compounds must either be volatile or capable of derivatization. Non-volatile matrices (wafers, oils, metal parts, etc.) require additional preparation (extraction, outgassing, etc.). Atmospheric gases are challenging ($CO_2$, $N_2$, $O_2$, Ar, CO, $H_2O$)

# Chapter 6 - Ion Exchange Chromatography (IEC)

## 6.1 Introduction

Ion-exchange chromatography (IEC) is part of ion chromatography which is an important analytical technique for the separation of inorganic ions, both cations and anions. Separation is based on ionic interactions between the charged species and ionic functional groups fixed on a column matrix having an opposite charge. Ion exchange chromatographic techniques used for the separation and purification of amino acid proteins, polypeptides, nucleic acid and other charged biomolecules.

## 6.2 Matrix used

An ion exchanger consist of an insoluble matrix to which charged functional group have been covalently bound. These charged groups are involved in exchange process that attract the opposite charged mobile ion and form an ionic bond with it. The column matrix is based on inorganic compounds, synthetic resin or polysaccharides. The characteristic of the matrix is based on its application. Application such as water treatment and recovery of ions from wastes such ion exchanger consist of hydrophobic polymer matrix. But the application is based on biological material, hydrophilic agarose and cross linked cellulose matrix is used.

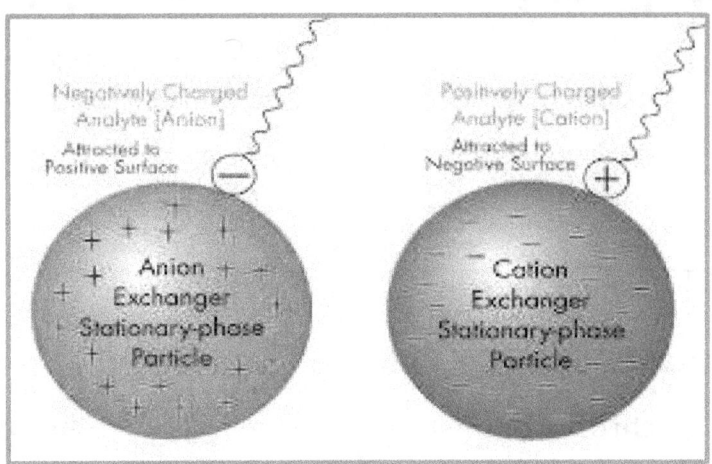

**Figure 6.1 Schematic diagram of cation and anion exchangers**

### 6.2.1 Cation Exchanger

The resin contain acidic functional groups aromatic ring of the resin. They have negatively charged functional groups and separate positively charged species. The strong-acid cation exchangers have sulfonic acid ($-SO_3H$) groups. The weak-acid cation exchanger have carboxylic acid group. The protons on these groups can exchanged with other cations:

$$nRz\text{-}SO_3^-\text{---}H^+ + M^{n+} \leftrightarrows (Rz\text{-}SO_3)_n M + nH^+ \quad \ldots\ldots(6.1)$$

and

$$nRz\text{-}CO_2^-\text{---}H^+ + M^{n+} \leftrightarrows (Rz\text{-}CO_2)_n M + nH^+ \quad \ldots\ldots(6.2)$$

Where Rz represent the resin. The equilibration can be shifted to the left or right by increasing $[H^+]$ or increasing $[M^{n+}]$ or decreasing one with respect to the amount of resin present.

The terms strong and weak cation exchanger refer to extent of variation of ionization with pH and not the strength of binding. Weak acid cation exchanger resin are more restricted in the pH rage (5 to 14), while the strong acid resins can be used from pH 1 to 14. The ion exchange capacity affects solute retention. The exchangers with high exchange capacity are most often used for separating the complex mixtures.

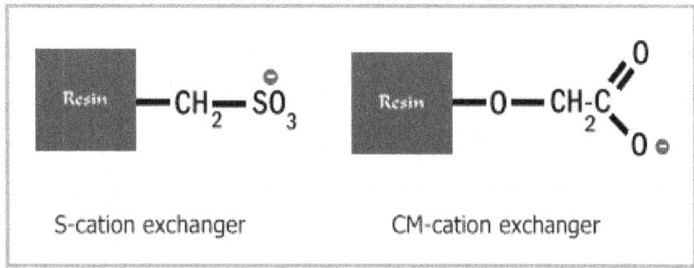

Figure 6.2 Schematic diagram of cation exchangers

### 6.2.2 Anion Exchanger

The resin contain basic functional groups. They have positively charged functional groups and separate negatively charged species. The strong-base exchangers have quaternary ammonium ($-NR_3$) groups and weak-base exchanger have amine group. The protons on these groups can exchange with other cations:

$$nRz\text{-}NR_3^+ \text{---}OH^+ + A^{n-} \leftrightarrows (Rz\text{-}NR_3)_n A + nOH^- \quad \ldots\ldots(6.3)$$

and

$$nRz\text{-}NH_3^+ \text{---}OH^+ + A^{n-} \leftrightarrows (Rz\text{-}NH_3)_n A + nOH^- \quad \ldots\ldots(6.4)$$

Where R represents organic groups, usually methyl. Weak base exchanger resin are can be used over the pH range 0 to 12, but the weak-base exchangers only over the range of 0 to 9.

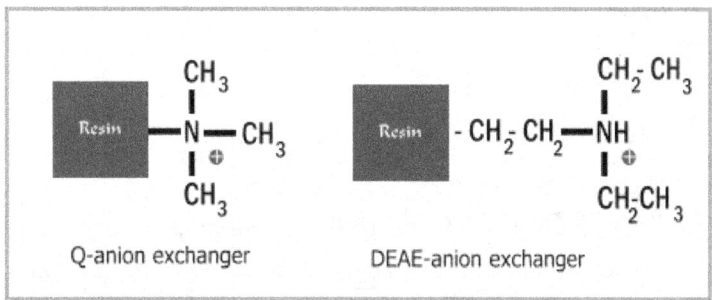

Figure 6.3 Schematic diagram of cation exchangers

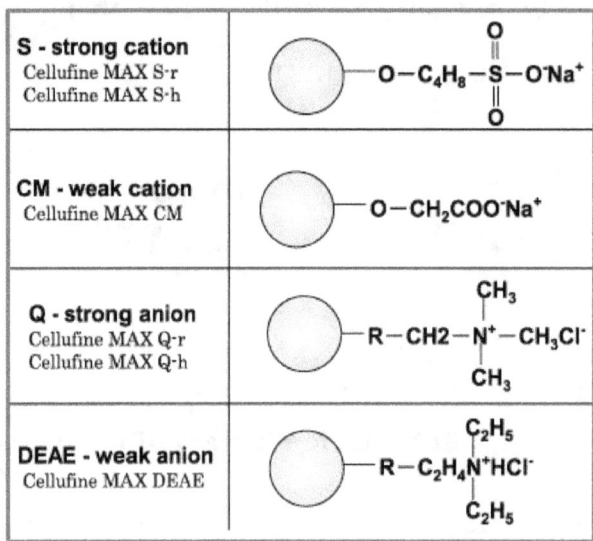

Figure 6.4 Types of ion exchange resins

## 6.3 Principal

Ion-exchange chromatography separates molecules based on their respective charged groups and analyte molecules are retained on the column based on ionic interactions. The stationary phase contains charged functional groups. The solute molecules are retained or eluted based on their charge. Initially, molecules that do not bind or bind weakly to the stationary phase are first to wash away. Altered conditions (increase a concentration of sample solution or change the pH of solution) are needed for the elution of the molecules that bind to the stationary phase. The ionic compound consisting of the cationic species $M^+$ and the anionic species $B^-$ can be retained by the stationary phase.

Cation exchange chromatography retains positively charged cations because the stationary phase displays a negatively charged functional group:

$$R\text{-}X\text{-}C^+ + M^+B^- \rightleftharpoons R\text{-}X\text{-}M^+ + C^+ + B^- \quad \ldots\ldots (6.5)$$

Anion exchange chromatography retains anions using positively charged functional group:

$$R\text{-}X^+A^- + M^+B^- \rightleftharpoons R\text{-}X^+B^- + M^+ + A^- \quad \ldots\ldots (6.6)$$

Note that the ion strength of either $C^+$ or $A^-$ in the mobile phase can be adjusted to shift the equilibrium position, thus retention time.

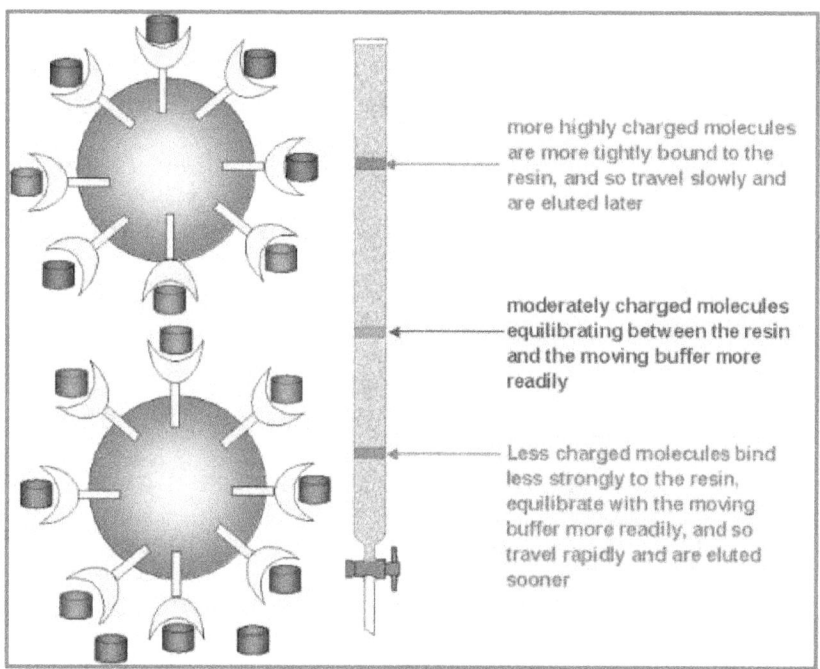

**Figure 6.4 Mechanism of ion exchange resins**

## 6.4 Steps in an Ion exchange Separation

An ion exchange process, medium contain a spherical particles substituted with ionic groups that are negatively or positively charged. The matrix is usually porous to give a high internal surface area. The medium is packed into a column to form a packed area. The bed is then equilibrated with buffer which fills the pores of matrix and space in between the particles.

### a) Equilibration

The first step is the equilibration in which the ion exchanger is equilibrated with desired start buffer solution. The pH and ionic strength of the start buffer solution allows to bind the desired solute molecules. When equilibrium is reached, all stationary phase charged groups are able to bind with exchangeable counter ions, such as chloride or sodium.

### b) Sample application and Adsorption

The second step is sample application and adsorption in which solute molecules carrying the appropriate charge displace counter ions and bind reversibly. Unbound substances van be washed out from the exchanger bed using starting buffer.

### c) Elution

In third stage, substances are removed from the column by changing to elution conditions that normally involves increasing the ionic strength of eluting buffer or changing its pH. Desorption is achieved by introducing of an increasing concentration of sample and solute molecules are released from the column in the order of their binding strength. The most weekly bound substances being eluted first.

### .d) Regeneration

A final wash with high ionic strength buffer regenerates the column and removes any molecules still bound. This ensures that the full capacity of the stationary phase is available for the next run. The column is then re-equilibrated in start buffer before starting the next run.

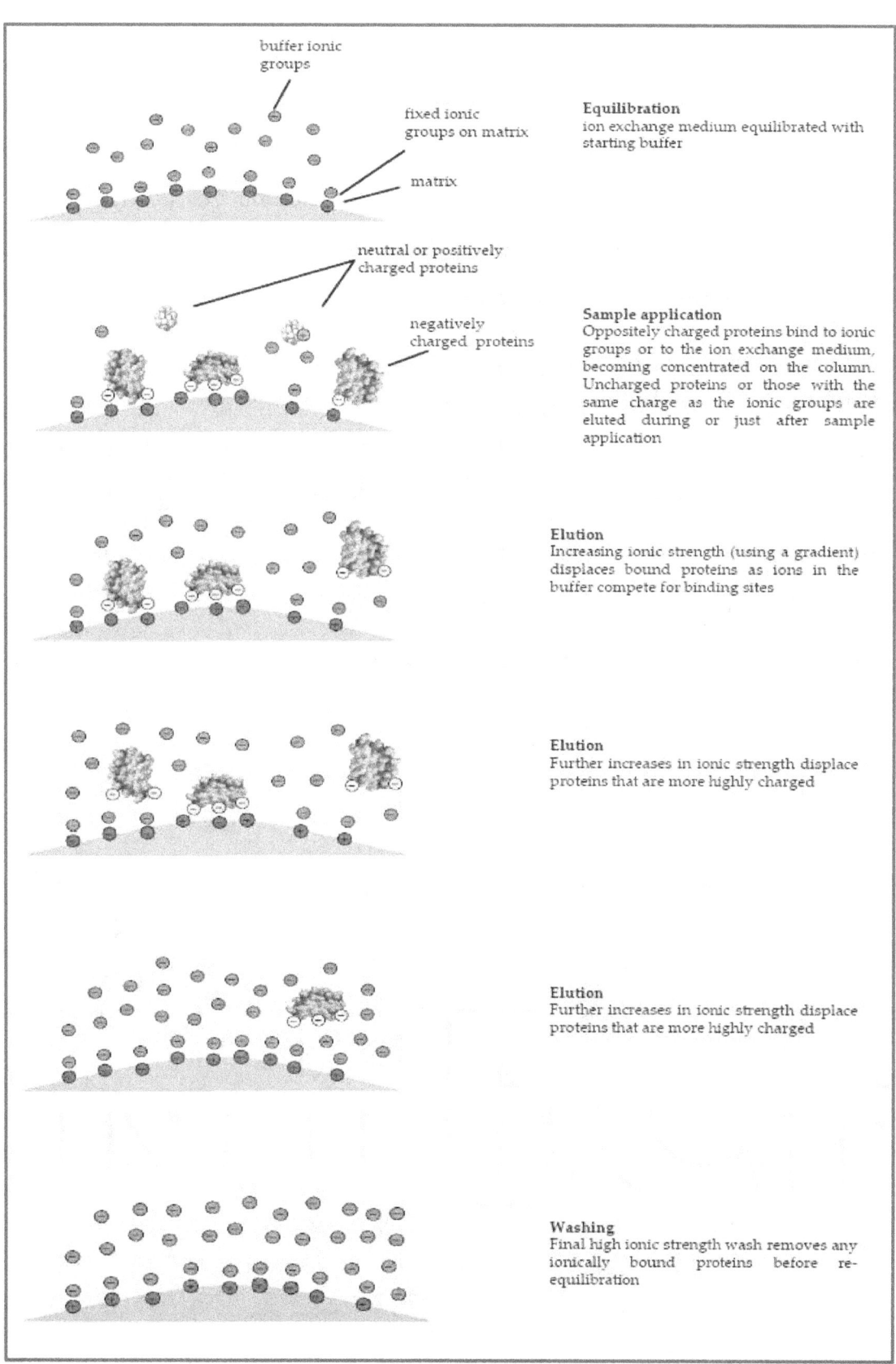

**Figure 6.5 Steps involve in ion exchange Separation**

## 6.5 Applications

### 1) Separation of Proteins using cation exchanger

Ion exchange chromatography is widely used for protein purification because most proteins carry zero net electrostatic charges at all pH except at pH=pI (isoelectric point). At a pH>pI of a given protein, that protein becomes negatively charged (an anion), at the pH<pI of that same protein, it becomes positively charged (a cation). The exchange process occurs due to electrostatic attraction between buffer-dissolved charged proteins and oppositely charged binding on the column which is pre-equilibrated with the buffer of identical pH and similar ionic strength as protein mixture. Protein mixture is applied onto the column. Positively charged proteins absorb to the media, displacing sodium cations. Retained proteins are eluted from the column with increasing salts concentration. To avoid drastic change of salt gradient (< 1M), pH (decrease) in column and applied concentration of adsorbing proteins (< 5 mg/ml).

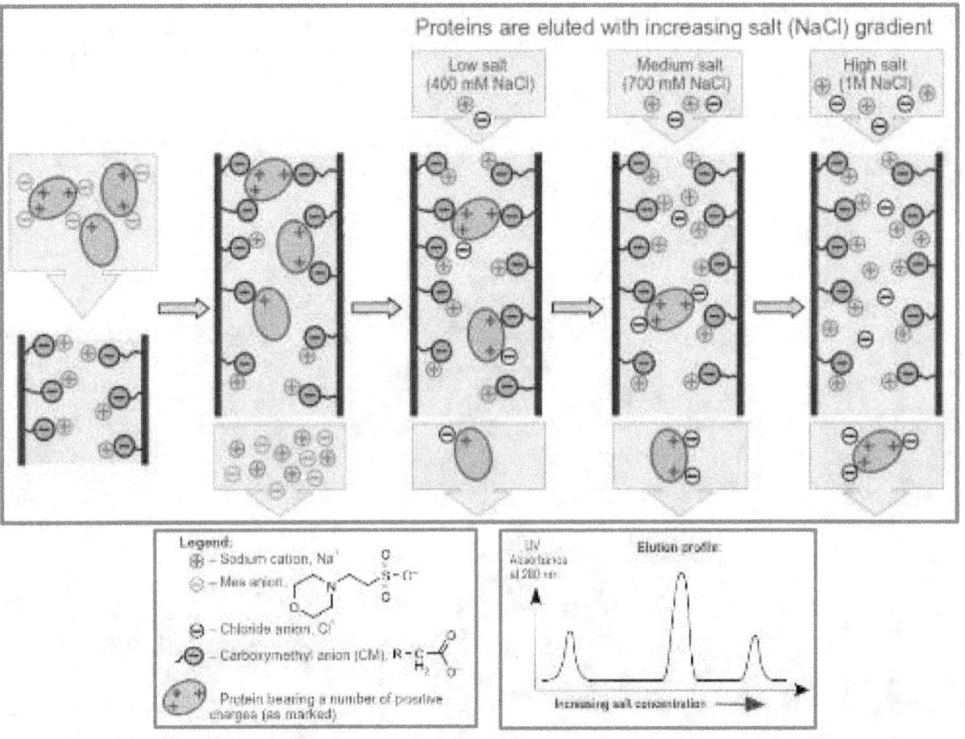

**Figure 6.6 Separation of proteins with cation exchanger**

## 2) Softing of hard water

When a hard water passed through cation exchanger which removes all cations a like $Ca^{+2}$, $Mg^{+2}$ etc. from it and equivalent amount of H+ ions are released from this exchanger to water. Thus water received from cation exchanger is an acidic in nature.

$$RH_2 + CaCl_2 \longrightarrow RCa + 2\ HCl$$

$$RH_2 + MgSO_4 \longrightarrow RMg + 2\ H_2SO_4$$

Acidic water is passed through an anion exchanger which removes all the anions like $SO_4^{-2}$, $Cl^-$, $NO_3^-$ etc. present in the water and release amount of $OH^-$ from this exchanger to water.

$$R'(OH_2) + 2\ HCl \longrightarrow R'Cl_2 + 2\ H_2O$$

$$R'(OH_2) + H_2SO_4 \longrightarrow R'SO_4^- + 2\ H_2O$$

**Regeneration:** Exhausted cation exchanger is regenerated by using dil HCl. and exhausted anion exchanger is regenerated by using dil NaOH.

$$RCa + 2\ HCl \longrightarrow RH_2 + CaCl_2$$

$$RMg + 2\ HCl \longrightarrow RH_2 + MgSO_4$$

$$R'Cl_2 + 2\ NaOH \longrightarrow R'(OH_2) + 2\ NaCl$$

$$R'SO_4 + 2\ NaOH \longrightarrow R'(OH_2) + Na_2SO_4$$

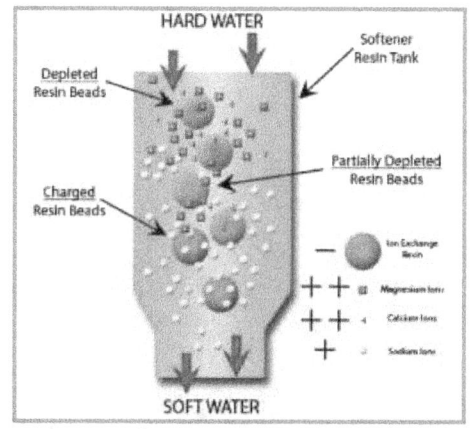

**Figure 6.6 Softing of hard water with ion exchangers**

# References

1. **Fundamentals of Analytical Chemistry** by Skoog, West, Holler and Crouch, 2006- Indian Edition (8th Ed) Pub: Thomson, Brooks/Cole
2. **Principles of Instrumental Analysis** by Skoog, Holler and Nieman, 2006- Indian Edition Pub: Thomson, Brooks/Cole
3. **Instrumental methods of Chemical Analysis** by B K Sharma (21st Ed), 2002; Goel Publishing House, Meerut
4. **Basic Concepts of Analytical Chemistry** By S M Khopkar (3rd Ed), 2008 New Age International (P) Limited, New Delhi
5. **Analytical Chemistry** by Gary D. Christin (6th Ed)

www.ingramcontent.com/pod-product-compliance
Lightning Source LLC
Chambersburg PA
CBHW080947170526
45158CB00008B/2401